Ana Brasil Machado

Ecolimits in Rio de Janeiro

AF570881

Ana Brasil Machado

Ecolimits in Rio de Janeiro

Policy and management of urban interfaces

ScienciaScripts

Imprint
Any brand names and product names mentioned in this book are subject to trademark, brand or patent protection and are trademarks or registered trademarks of their respective holders. The use of brand names, product names, common names, trade names, product descriptions etc. even without a particular marking in this work is in no way to be construed to mean that such names may be regarded as unrestricted in respect of trademark and brand protection legislation and could thus be used by anyone.

Cover image: www.ingimage.com

This book is a translation from the original published under ISBN 978-3-330-73751-8.

Publisher:
Sciencia Scripts
is a trademark of
Dodo Books Indian Ocean Ltd. and OmniScriptum S.R.L publishing group

120 High Road, East Finchley, London, N2 9ED, United Kingdom
Str. Armeneasca 28/1, office 1, Chisinau MD-2012, Republic of Moldova, Europe
Printed at: see last page
ISBN: 978-620-8-26871-8

Copyright © Ana Brasil Machado
Copyright © 2024 Dodo Books Indian Ocean Ltd. and OmniScriptum S.R.L publishing group

Summary

ACKNOWLEDGMENTS

As this book is a version of my Master's thesis, I would like to thank those who helped develop it.

To CAPES, for the financial aid granted during my Master's Degree in Geography at the Federal University of Rio de Janeiro.

To Prof. Dr. Paulo Cesar da Costa Gomes, my advisor, for his tolerance and for teaching me, every day, the enormous love of intellectual work.

To Prof. Ms. Leticia Parente Ribeiro, who dedicated countless hours of her time and made this work a pleasant and precious dialogue.

To Prof. Dr.ª . Ana Marcela Ardila, for her contributions in the early stages of writing.

To the members of the defense and qualifying exam boards: Prof. Dr. Frederico Araujo, for his willingness to participate and for gifting the world with his inventiveness and rigor; Prof. Dr. Rafael Winter Ribeiro, for his willingness to participate and for following this work from the very beginning, and Prof. Dr. Paulo Gusmao, for his generous contributions to the Qualifying Exam.

To my colleagues in the Territory and Citizenship Research Group, for the cheerful and cooperative working environment, especially Andre Felix, Nikolas Zanette and Paula Trojan.

To my family and friends who have made my journey easier and happier. In particular, I would like to thank Lili and Zela, who have always held my hand, and Felipe, who shared the adventure of seeing a new life being born during the Master's course.

To Flora, who made life so important.

Introduction

The management of urban sprawl has historically been the subject of numerous forms of government intervention, which have taken the form of urban policy instruments such as wall systems, zoning, green belts, urban service zones, building taxes, among others. In some cases, these instruments took the form of physical boundaries, reinforcing or creating internal and external differentiations within the city.

Physical demarcation was widely used in ancient and medieval cities. Many of them retain, in their current morphologies, traces of ancient walls or real wall systems that performed a variety of functions, such as protection against enemy peoples and wild animals, the collection of taxes and the demarcation of areas with administrative functions (CAPEL, 2002).

From a geographical point of view, it's important to answer a few questions: how are boundaries and discontinuities in cities produced, materially and symbolically? Which spatial categories are mobilized in policies to contain urban sprawl? What are the interfaces between these categories? By studying the construction of the Ecolimites public policy in Rio de Janeiro, we seek to understand these issues in a specific urban context.

Throughout the 2000s, in the city of Rio de Janeiro, the notion of ecolimits was used at various times as an instrument for managing urban growth within the city. These were physical delimiters that should be located at the contact between favelas and environmental preservation areas. In 2001, the project was conceived and implemented by the City Hall, through the Municipal Department of the Environment, and consisted of concrete markers interconnected by steel cables that were installed along the lines of contact between the favelas and the areas to be preserved in more than thirty locations.

In 2004, the then deputy governor proposed the construction of walls that were supposed to contain the growth of the favelas and hinder the actions of drug traffickers who used the surrounding forests as escape routes. Even though they weren't implemented at the time, the walls were able to mobilize a debate about their probable inefficiency and the meanings they carried. According to Compans,

> "In September 2005, the newspaper *O Globo* began a series of reports focusing on the City Hall's omission in the face of the disorganized growth of Rio's favelas. In fact, this issue had already been covered by the press in April 2004, when a war erupted over control of drug trafficking in Rocinha, terrorizing the residents of the city's south zone. At that time, there was an intense debate in the media about possible solutions to contain the expansion of the favelas, including the State Government's proposal, presented by the Vice-Governor and State Secretary for the Environment and Urban Development Luiz Paulo Conde, Luiz Paulo Conde, which consisted of surrounding four of them - Rocinha, Vidigal, Parque da Cidade and Chàcara do Céu - with a wall three meters high, and developing a "social occupation" in them, combining ostentatious policing and medical and dental care" (COMPANS, 2007: 88).

In 2008, a group of municipal secretaries announced the construction of more "robust" ecolimits than the old ones. They would take the form of three-meter-high walls and would be installed in at least eleven favelas located in the South Zone of the city: Dona Marta, Rocinha, Chapéu Mangueira, Benjamin Constant, Chacara do Céu, Parque da Cidade, Morro dos Cabritos, Tabajaras, Pavào-Pavàozinho, Cantagalo and Vidigal. The project's stated aim was to contain the growth of favelas over the city's green areas. According to icaro Moreno, president of the Public Works Company of the State of Rio de Janeiro (EMOP), "the idea is to protect the community on one side and the Atlantic Forest on the other. It makes monitoring easier and helps to scale the actions in the communities" (O GLOBO, 28/03/09).

The first of the new ecolimits was built in the Santa Marta (or Dona Marta) favela in Botafogo in 2009. In this favela, a concrete wall about 600 meters long was built, running from the top of the hill to station 2 of the inclined plane, where it coincides with the wall of the Palacio da Cidade do Rio de Janeiro, belonging to the City Hall .[1]

The implementation of this ecolimit was the subject of intense public debate, which was explored in a previous work (MACHADO, 2009). Newspapers and magazines devoted significant space to authors with different opinions, official *websites* such as EMOP's published articles, and *blogs* organized by Santa Marta residents also published texts about the construction of the walls. What was in question was the very validity and necessity of building these boundaries. The objective of containing the expansion of the favelas over the green areas was questioned, since the first favela to receive the project showed a decrease in its area between 1999 and 2004, as described by Cavallieri and Lopes (2006).

However, most of the discussion focused on the formal issue of the boundaries: their materiality, their wall-like form, produced various reactions that almost always brought up the great symbolic charge of this type of object. This public debate brought into play various values and social practices related to life in the city, the favela and environmental preservation areas. The boundary thus took on different meanings, establishing spatial relationships between formality and informality, nature and artifice in the city. These meanings resulting from the debate were concentrated around three main axes: segregation, order/disorder and environmental preservation (MACHADO, 2011).

So intense were the debates that the project created for the Rocinha ecolimit, the second favela to receive the project, had to be revised and negotiated with local leaders and other sectors of society, so that the three-meter wall gained alternatives such as low walls and fences or was associated, as in

1 According to EMOP officials, there has been a change to the original layout of the project. The wall, which was supposed to follow the water channel, was diverted to increase the area of the favela so that it could house four apartment buildings for the families removed from the upper part of the hill.

the project implemented, with an ecological park with various facilities for use by the favela's residents.

The Rocinha ecolimit began to materialize in 2010, but work was suspended in 2011 and resumed at the beginning of 2012. In fact, only Santa Marta and Rocinha received the project's works, despite the goal of building a total of eleven kilometers of walls. However, the small area built was able to mobilize various discourses about the city. The apparent emptying of materiality gave way to a strong dispute in the field of ideas and returned as an object of tension. Simply stating the materialization of a boundary or a wall in a given place has the power to mobilize public debate, to produce normative instruments and urban planning projects, to summon up spatial categories, and to concentrate structuring issues in the city of Rio de Janeiro.

The debate also reached the legislature of the city of Rio de Janeiro, culminating in the formulation of a bill which, in regulating ecolimits, intended that they could not be materialized in the form of walls. Ecolimits were also debated in the context of the Rio de Janeiro Master Plan. The evolution of this instrument can be followed if we consider the different documents produced over the years in an attempt to revise the 1992 Master Plan .[2]

Some academic papers have been produced on the ecolimits project since 2009. Most of them identify this public policy with the issue of socio-spatial segregation and claim that the discourse of environmental preservation is just an instrument used to mask the state's real intentions. Despite the relevance of the analyses produced, we believe that this body of work is based on an explanatory grid that is presented before the phenomenon is even observed.

In this research we propose that, beyond the problem of segregation, ecolimits act fundamentally in the management of urban growth. Taking into account the specificities of the city of Rio de Janeiro, where the contact between urbanized space and areas of environmental protection is especially valued, ecolimits produce and requalify spatial cut-outs and discontinuities within the city itself.

Given the centrality of the issue for the city, ecolimits are not an isolated instrument: many other instruments also act to control urban growth. These are presented as antecedents, conditions and elements with which the ecolimits project is confronted in the formation of a device that responds to the urgency attributed to the growth of slums.

---

2 According to the Cities Statute (2001), the master plans of cities with more than 20,000 inhabitants or which are part of metropolitan regions or urban agglomerations must be revised every ten years. The revision of Rio de Janeiro's Master Plan, however, was only published in 2011.

1 Urban expansion and regulatory instruments in Rio de Janeiro

Academic works on ecolimits almost unanimously link the construction of the project's walls to the issue of socio-spatial segregation. Far from denying the importance of the debate on the problem that is undoubtedly present in the city of Rio de Janeiro, it is believed that the formation of a consensus in the interpretation of the phenomenon of ecolimits contributes little to its understanding. Ecolimits seem to concentrate in their formulation, implementation and discussion a series of issues pertinent to the city: relations between the formal city and the informal city, the preservation of forest fragments in an urban environment and the control of the growth of urban occupation in specific areas.

In this sense, we propose that ecolimits are also instruments for managing city growth, acting as physical delimiters. Various instruments are used for this purpose, such as zoning, green belts, urban service zones, building taxes and wall systems, which were used in ancient and medieval cities. With the exception of inner walls and citadels, as described by Capel (2002) and ecolimits, these instruments for managing urban growth were designed to contain expansion towards the outer limits of the city, towards the peripheral zones. This "outward" growth was and is the concern of a large number of cities such as London, Seoul and Sao Paulo, which have set up green belts.

In the case of the city of Rio de Janeiro, until the 1960s, population growth was higher than the national average, due to intense migration to the city. From then on, the rate of growth began to decline more rapidly than in the country as a whole (IPP, 2004a). However, this growth has not been evenly distributed throughout the city. Different parts of the city can concentrate a larger population, have a higher density and be characterized as areas of city expansion or as consolidated areas, demonstrating the dynamic nature of the city and its growth.

For planning purposes, the city of Rio de Janeiro was divided into five large areas, as can be seen in the figure below.

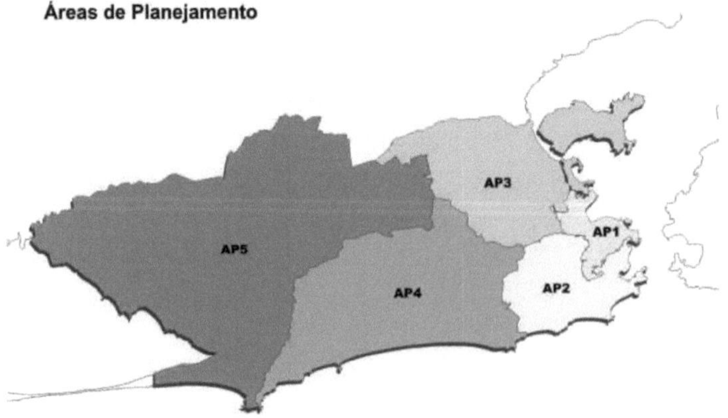

Figure 1: Rio de Janeiro City Planning Areas

Source: Rio de Janeiro City Hall

Since 1980, the population of Planning Area 1 (PA 1) has been decreasing, and in 2000 it represented less than 5% of the municipality's total population. Planning Area 2 (PA 2) and Planning Area 3 (PA 3) have also been losing population, accounting for 17% and 40% respectively in 2000. Planning Area 4 (PA 4) and Planning Area 5 (PA 5) are areas where the population has been growing since 1980. PA 4 accounts for 11.6% and PA 5 for 26.6% of the total population of the municipality of Rio de Janeiro (CAMARANO et al, 2004).

According to a study carried out by the IPP (OLIVEIRA, 2008), these last two planning areas are in the process of consolidation and are characterized as areas of urban expansion. While PA 1 and PA 2[3] participated negatively in the total population increase of the municipality, PA 4 participated with 42.5% (1980 to 1990) and 41.5% (1990 to 2000) and PA 5 presented, for the same periods, a participation of 68.1% and 69.5%. PA 3 contributed 19.4% in the period 1980 to 1990 and 9.5% in the period 1990 to 2000 to the total population growth in the municipality of Rio de Janeiro.

The municipal government can contribute to this dynamic insofar as it establishes areas for the expansion and containment of the urban fabric, promotes infrastructure works and popular housing programs. The Sustainable Urban Development Master Plan for the Municipality of Rio de Janeiro (Complementary Law 111/2011), Rio de Janeiro's main urban planning instrument, has urban expansion as one of its guidelines (Article 3).

Article 32, in the chapter dedicated to Macrozoning, establishes four macrozones, which can be seen in the figure below.

3 For the period from 1980 to 1990, AP 1 had a share of -8.4% and AP 2 of -21.6%. For the following period (1990-2000), AP 1 had a share of -9.9% and AP 2 - 10.6% (OLIVEIRA, 2008).

Figure 2: Macrozones established by Rio de Janeiro's Master Plan

Source: Sustainable Urban Development Master Plan for the Municipality of Rio de Janeiro, 2011

In the Controlled Occupation Macrozone, which covers Planning Areas 1 and 2, population density and construction intensity must be limited. The Conditional Occupation Macrozone, which encompasses part of PAs 4 and 5, makes expansion conditional on the capacity of existing infrastructure or that dependent on private capital, and on environmental and landscape protection. In the Incentivized Occupation Macrozone, where parts of PAs 3, 4 and 5 are located,

"Population densification, construction intensity and the increase in economic activities and large-scale facilities will be encouraged, preferably in areas with greater availability or potential for infrastructure deployment" (RIO DE JANEIRO, 2011).

The Assisted Occupation Macrozone coincides with the Campo Grande and Santa Cruz Administrative Regions, both located in AP 5,

"where population densification, increased economic activity and the installation of economic complexes must be accompanied by public investment in infrastructure and measures to protect the environment and agricultural activity" (RIO DE JANEIRO, 2011).

Despite urban planning efforts, Rio de Janeiro's rapid demographic growth and expansion were characterized by the production of a city "far from the public realm" (MAGALHÂES, 2012: 181-182). The favelas are spread all over the city, even in more affluent areas such as the South Zone (AP 2), and the occupation of the hillsides is a hallmark of Rio's landscape.

"As early as the 1920s and 1930s, they were considered "stains" on the city's urban landscape and were invariably recommended for extinction. Urban plans and regulations (including the Agache plan, finalized in 1930) placed them outside urban legality and indicated their removal under arguments inspired mainly by sanitation, but also by functional and aesthetic aspects" (CARDOSO and ARAUJO, 2007: 278).

For these authors, in the 1960s, during Carlos Lacerda's government, the aim of eradicating the favelas led to the demolition of shacks and removal to the Vila Kennedy and Vila Aliança housing estates, located in the West Zone. This decade was also marked by proposals to urbanize the favelas and, therefore, by the polarization of the debate between removal and urbanization. In the 1970s, this polarization intensified, on the one hand due to the BNH (National Housing Bank) evictions and, on the other, due to the international debate on housing issues, which was marked by the First International Habitat Conference.

During the 1980s, the same BNH worked on the urbanization of the Maré favela and the municipal and state governments began to develop their own projects for the favelas. Leonel Brizola's government was responsible for two pilot projects for favela urbanization, "laying the foundations for an intervention methodology that would be perfected in the following years". These projects were based on the ideas of concentrating interventions on infrastructure rather than housing, promoting accessibility, and creating social and leisure facilities in the favelas (CARDOSO and ARAUJO, 2007: 280). In the 1990s and 2000s, projects such as Favela-Bairro and Bairrinho consolidated the commitment to favela urbanization rather than evictions, even though evictions were carried out or planned in different situations, such as the cases of Parque Rebouças and Vila Autòdromo, despite the ban established in the 1992 Master Plan .[4]

According to a technical note from the Pereira Passos Institute (BESSERMAN and CAVALLIERI, 2004), the population growth rate in the city of Rio de Janeiro was 0.74% in the period from 1991 to 2000. For the same period, favela and non-favela areas showed different rates: 2.22% and 0.43%, respectively, this difference being credited to the higher migration and fertility rates in the favela.

With regard to the areas occupied by favelas between 1999 and 2004, the Pereira Institute

Passos (CAVALLIERI and LOPES, 2006) carried out a visual analysis of orthophotos (taken in 1999 and 2004) of 750 favelas. The results only refer to the horizontal expansion of the favelas, and it was not possible, using the method employed, to analyze the vertical growth and densification of the plots.

For the city as a whole, there was a 3.5% increase in the area occupied by slums.

While PAs 1, 3, 4 and 5 showed an increase in slum area over the period covered by the study, PA 2, where the two built ecolimits are located, showed a decrease of 0.2%. In Santa Marta, the favela where the first wall of the 2009 project was built, there was a greater decrease than the AP average (-0.6%). Rocinha, on the other hand, saw an increase in its area of 1.4% over the period considered, a percentage higher than that of AP 2 and lower than that of the city (CAVALLIERI and LOPES, 2006).

Because the genesis of the carioca favela was in the hills and this is the recurring topography of highly visible favelas, the word hill is often used as a synonym for favela. And, in many cases, favelas are in areas set aside for environmental protection or in *non-aedificandi* areas, such as Permanent Preservation Areas (APP), established by the Forest Code (1965) (repealed by Federal Law 12.651 of 2012), areas of forest or native vegetation on slopes with a gradient greater than 45 degrees and hilltops, for example. In this way, the discussion about environmental preservation in Rio necessarily involves the issue of containing urban occupation on the slopes and, more specifically, containing the

---

4 "(...) One of the principles of the City's Master Plan, approved in 1992, is the non-removal of favelas and their transformation into neighborhoods, through financial and urban regularization and the provision of infrastructure and urban facilities. However, the principle of non-removal does not apply to slums that occupy a) conservation units or areas of special environmental interest; b) risk areas; c) marginal strips protecting surface water, water mains and high-voltage electricity grids; d) strips of federal, state and municipal roads; e) spans and pillars of viaducts, bridges and footbridges, as well as adjacent areas, when they pose a risk to individual and collective safety or make the implementation of basic urban services unfeasible; f) or areas that cannot be provided with minimum conditions of modernization and basic sanitation" (COMPANS, 2007. 85): 85).

growth of favelas on green areas and hillsides. According to Rose Compans:

> "It must be noted that, while at first, by defining hillsides as areas to be protected, environmental legislation ended up facilitating their occupation by the poor, given the lack of interest from the real estate market, it has now provided a new justification for the containment or even removal of these informal settlements. In addition to the administrative delimitation of environmental conservation units, there has been a widespread perception - legitimized by technical-scientific discourse - that the favela constitutes a risk to the community, either because of the possibility of natural disasters or because of the characteristics of the occupation - such as the lack of sanitation and the high population density - which are factors in the degradation of the urban environment" (COMPANS, 2007: 84).

In addition to the Brazilian Forestry Code, other instruments are used to manage the expansion of the urban fabric in the city of Rio de Janeiro. This is the case of the 1976 Construction Code (Decree 322), which, for example, establishes altimetric levels above which construction is prohibited.

> "In the 1970s, to the extent that it was (re)known, still embryonically, that forest shrinkage was contributing to making the soils on the slopes less stable and that this process would increase the risk and frequency of landslides and floods in the lowlands surrounding the city's massifs and hills - as indeed happened in emblematic fashion in 1966 and 1967 - the urban legislation of the former state of Guanabara instituted the first more effective restrictions on occupation. the urban planning legislation of the former state of Guanabara instituted the first more effective restrictions on the formal occupation of the slopes of the then Federal District (municipal decrees 3800/1970 and 6168/1973 and later municipal decree 322/1976)" (SCHLEE and ALBERNAZ, 2009 :5).

Decrees such as 2.638, of 1980, and 3.155, of 1981, amend the Zoning Regulations approved by Decree 322. Decree "E" 6.168, of 1973, establishes restrictions on the opening of streets on hillsides. Decree 8.321, on the other hand, establishes conditions for buildings on slopes located in the Residential 1 and Special 1 Zones.

We can consider that the creation of Areas of Special Environmental Interest (AEIA) and Environmental Conservation Units (such as the Environmental Preservation Area - APA, Environmental Preservation and Urban Recovery Area - APARU, Area of Relevant Ecological Interest - ARIE, Biological Reserve - REBIO and Parks) acts decisively in directing and containing urban expansion. This is also the case with management plans, such as the 2008 Tijuca National Park Management Plan, which proposes the creation of its Buffer Zone, an area that aims to minimize edge effects by restricting uses and occupations around conservation units.

As already mentioned, the Master Plan also interferes in urban expansion insofar as, for example, it establishes zones where occupation must be controlled and zones where the government must invest in the creation of infrastructure, making new occupation possible. Article 2, paragraph 1 states that:

> "Urban occupation is conditioned to the preservation of massifs and hills; forests and other areas with vegetation cover; the seafront and its restinga vegetation; water bodies, lagoon complexes and their marginal strips; mangroves; landmarks and the City's landscape" (RIO DE JANEIRO, 2011).

The third and fourth paragraphs of Article 2 deal with the city's landscape. This is understood by the document as the spatial configuration resulting from the interaction between the natural and cultural environments and considered "the most valuable asset of the City, responsible for its consecration as a world icon and for its inclusion in the country's tourist economy, generating employment and income" (RIO DE JANEIRO, 2011). According to Magalhaes,

"Today, the 'return to green' in big cities is nothing more than the flight of segments of the upper middle class to the outskirts, where they live isolated on generous plots in gated communities, in a reduced version of the North American suburb. However, Rio de Janeiro has a different relationship, where in the theater of urban life the elements of mountain, forest and water take on the role of protagonists, not supporting actors. In Rio, nature is not the 'absence of the urban' or the 'emptiness of occupation': on the contrary, nature is the urban. The beach is the best example. Nature is occupation: mountains and forests that define spaces" (MAGALHÀES, 2002: 23).

In this way, the presence of forests, beaches and massifs contributes to shaping the city of Rio de Janeiro not only in the sense of providing environmental services. The juxtaposition of natural elements and the urban fabric is a hallmark of the city, producing a strong image that identifies it. This "harmonious encounter" enabled UNESCO to award the city the title of World Heritage Site in the Cultural Landscape category in 2012.

All these tools point to the importance of controlling the expansion of the city, which in Rio is not limited to growth towards the outskirts. The problem with growth in Rio is not that of sprawling expansion, towards the peri-urban fringes, but rather growth within internal limits, towards specific areas within the urban zone, which must be maintained in order to guarantee the persistence of the city's image. It is in this context that the ecolimits policy was produced throughout the 2000s, going through different stages and adaptations. The construction of physical boundaries between the favelas and the green areas to be protected are necessarily related to the other instruments for managing urban growth. This relationship takes place abstractly, in laws and concepts, but also in the conflicts that take shape on the ground, when norms and other apparatuses are applied.

2. On the fence: questions, objectives and hypotheses

The more general question we seek to answer here can be formulated as follows: *What was the process of formulating and implementing the public policy known as "ecolimits" in the city of Rio de Janeiro like?*

From this question, other questions emerge, such as: what are the objectives guiding the construction of the ecolimits? What are the morphologies mobilized by the project? Who are the agents involved in its formulation and implementation process? What are the spatial categories placed in relation by the boundaries produced in the instruments/documents considered? What are the interfaces produced between these categories?

The general aim of this research is to analyze the process of drawing up and implementing the ecolimits policy in the city of Rio de Janeiro. This will be possible by identifying and describing the following elements:

a) The instruments that make up the ecolimits policy (legislation, technical studies, hearings, projects, workshops, etc., considering that this policy is made up of a certain set of instruments);

b) The contents of the documents that make up the policy in relation to the definition, location, objectives, and values attributed to the boundaries produced;

c) Those involved in the process of building and implementing the policy (executive and legislative bodies and their committees, architectural firms, associations, etc.);

d) The categories and spatial domains mobilized in the documents considered (favela, asphalt, green areas, occupied areas, places, etc.).

In this way, it will be possible to analyze the changes and continuities over time in the configuration of the ecolimits policy.

The hypotheses on which the work was carried out concern the component instruments of the ecolimits device[5], the contents of these documents, the agents who participated in the constitution of the ecolimits policy and the spatial categories mobilized in the production of interfaces and limits by the public policy documents to be analyzed (here also referred to as instruments and elements).

It is considered here that the ecolimits project is related to other issues of urban policy in the city of Rio de Janeiro and not just the problem of socio-spatial segregation. We intend to highlight the role of ecolimits as a policy that acts on urban growth by managing discontinuities within the city of Rio de Janeiro.

---

5 The concept of device, as discussed by Agamben (2011) on the basis of Foucault's ideas, will be discussed below.

We also believe that the ecolimits project is not only defined by its Bill of Law (PL 245/09) and specific urban designs. In this sense, we propose that the ecolimits present themselves as a *device* (in the sense used by Agamben, 2011) that relates elements of different natures, but which have in common the production of limits and interfaces between occupied or occupiable areas and areas of environmental protection, such as the Master Plan, the Construction Code and the laws that establish Conservation Units.

One of the central ideas that this work aims to develop is that city and nature are two domains that are requalified with the construction of a wall, a bike path, a park, a boundary. This requalification is linked to the location, morphology and spatial categories that the production of boundaries brings into play.

The ecolimits public policy should not be understood as an instrument impervious to disputes over interests and visions of the city. The production of this policy takes place as a non-linear process that involves negotiations with other agents and antecedents, where the final product is not finished and given as soon as the public authorities announce their intention to build walls in the favelas. What's more, the state, as the party responsible for formulating and implementing the project, is not a single, monocordial agent: what we call "the state" is made up of different levels of government and different bodies, which have different views on urban issues, environmental preservation and the instruments that can be used to solve the problem of urban sprawl.

A final working hypothesis concerns the spatial categories mobilized in the documents that establish the public policies related to the ecolimits project. In the different documents, produced by different agents and in different situations, with specific purposes and contexts, the categories placed at the interface of the establishment of limits are also different and present values about the city, nature and the act of limitation itself.

3. The debate on Rio de Janeiro's ecolimits

Disregarding the explicit objective of the 2008/2009 ecolimits project, namely to contain the urban growth of favelas on the green areas of the municipality[6] , academic works on the subject focus exclusively on the issue of socio-spatial segregation. These works have focused on the relationship between the formal city and the informal city, believing that the environmental preservation function of the boundaries is just an empty discourse that masks the real intentions of the state: to separate the favela populations from the rest of the city.

In this way, by applying an absolute explanatory grid to the investigation of the phenomenon in question, other meanings have not been explored which, however, could present other ideas of the city. Another common point among the works on ecolimits is a tendency to standardize the figure of the state. In general, the state is seen as a single, monolithic actor, producing a univocal speech, since it is considered to represent the elites in such a way as to act in favor of their economic interests, to the detriment of and contrary to the popular strata of society. This is not to say that the relationship between the formal city and the informal city, *the asphalt* and *the favela*, is not important or that it shouldn't be discussed, that it doesn't also organize life in the city. But it is to say that the notably spatial relationship between city and nature should also be problematized.

3.1. Ecolimits and socio-spatial segregation: building a consensus

In academic works about ecolimits or in works that have ecolimits as a sub-theme, it was possible to notice that the discourse on forest preservation contained in the project was considered to be an element that was not "true". The real intentions of the project, to build boundaries for the favelas so that they don't grow and therefore don't overlap with the green areas, are being masked by the "green discourse", which at the same time assumes a value that is very little questioned socially. In other words, for some authors, the ecolimits project uses a justification that is widely accepted by Brazilian and world society in order not to be questioned.

According to Pedroso, interventions in favelas at other times in history were based on sanitary arguments. The implementation of ecolimits, at the present time, appropriates an environmentalist discourse, appealing to a justification widely accepted in society. In this way, the very name of the project, the prefix "eco", is a strategy for social validation:

"(...) the discourse of the project's creators ratifies that the favelas are expanding horizontally over the municipality's Atlantic Forest reserves. Thus, even the way of referring to the walls has been adapted, since the language disseminated by the public authorities is ecolimits" (PEDROSO, 2010: 5).

---

6 According to Icaro Moreno, president of the Public Works Company of the State of Rio de Janeiro (EMOP), "the idea is to protect the community on one side and the Atlantic Forest on the other. It makes monitoring easier and helps to scale the actions in the communities" (O GLOBO, 28/03/09).

According to some of the works, this environmental discourse is just a façade, a disguise in the form of a speech, which hides the real intentions of the project. According to these authors, the environmentalist appeal would serve the state in the sense of "disguising" the elitist content and gaining "the sympathy of the public, especially the urban middle classes" (CAMARGO, 2012: 2):

"Obviously, it's not a question of seriously discussing whether the wall around Rio's favelas is really the right solution to stop the explosive and disorderly growth of Brazil's metropolises and protect the remaining natural areas. But to understand how a whole opportunistic and perverse political rhetoric appropriates contemporary consensual discourses and exploits the most deep-rooted national problems and prejudices in order to push through authoritarian, self-interested projects of questionable constitutionality in the name of democracy and environmental preservation" (CANÇADO, 2009: 3).

Haesbaert (2010), dealing more generally with walls, places the construction of ecolimits in a paradoxical context in which the imperatives of fluidity and avoidance clash. If, on the one hand, it is possible to talk about the opening up of borders, on the other, it is possible to see the "strengthening of an inverse process, that of a new proliferation of walls, fences or, if we wish, of territorial 'borders' in the broad sense" (HAESBAERT, 2010: 537). On the subject of ecolimits, Haesbaert states that the discourse of biosecurity, the type of discourse mobilized for the projection of limits and walls in the favelas, would try to hide other concerns of the state "with the 'risks' of drug trafficking and the very physical expansion of this mass of the poorest population, considered 'harmful' and a locus for generating insecurity, at least in the symbolic context of the asphalt-favela relationship" (HAESBAERT, 2010: 545).

The real content of the project, revealed despite state efforts to conceal it, is, according to a number of authors, that of strengthening the separation between the poor and the rich, the formal city and the favela, those inside and those outside: "It is an aggressive symbol that further increases the deep inequalities, now stigmatized, between those who live inside and those who live outside" (DE CARLI and HUMANES, 2010: 1)[7] . And its effects would unequivocally contribute to segregation: "If we take into account that on the other side of the favela there is an inclined plane, with a cable car to transport the community, which already serves as a containment wall, at the end of the construction of the wall the residents will be concretely segregated" (FERREIRA, 2009:15).

More than separating, for Cançado (2009), the ecolimits serve to immobilize and isolate those who are inside the walls, the favela residents. According to De Carli and Humanes, the aim of the project was to isolate the favelas from healthy parts of Rio de Janeiro. The result of this enclosure would inevitably be the formation of ghettos: "We understand that projects like these have an exclusionary and segregationist character, since they expose the residents of these places to a process of forced and

---

7 "We face an aggressive symbol that increases even more the deep inequalities, now stigmatized, among those that live inside and those that live outside." (DE CARLI and HUMANES, n.d.: 1).

imposed '*ghettoization*' (...)" (PEDROSO, 2010: 7).

In this sense, the ecolimits, or more specifically the wall built in Santa Marta, would be able to give visibility to other boundaries that already existed between the favela and the formal city even before the project was conceived. Still according to De Carli and Humanes, what had previously been invisible, "the walls of exclusion, injustice, omission and terror"[8] were legitimized by the Rio de Janeiro government, taking on a concrete appearance. The questioning of the effectiveness and legitimacy of the project led the authors of the report entitled "The Walls in the Favelas"[9] to believe that the limits would contribute even more to the formation of socio-limits by exacerbating the differences between the population living in the favelas and the middle classes living in the neighborhoods surrounding the favelas.

In the works cited so far, the environmental discourse hides the true interests of ecolimits. When they claim that the wall or the ecolimits are objects that create and/or reinforce socio-spatial segregation, the valuable opposition is that between the formal city and the informal city. All other meanings socially attributed to the project, or more specifically to the Santa Marta wall, are seen as fallacious, unimportant and elitist. According to this view, the environmental discourse is the strategy through which the real objective of the ecolimits is hidden. It would be up to these researchers and analysts to reveal, unveil and bring to light the real content of the project. From this point of view, the other meanings pointed out do not merit any exploration .[10]

Thus, the environmental discourse is much more taken as a mask, a fantasy that prevents us from getting closer to the object than a possibility for exploration, for questioning. However, the mobilization of the discourse of environmental protection also proposes an image of the city, it proposes a city. It puts words and actions into dispute, it composes relationships other than those marked only by the opposition between the formal city and the informal city.

3.2. The monolithic state and public policy as a "gift of deception"

Another aspect found in the works in question concerns the relationship between the state and the media. According to Pedroso (2010), the media, co-opted by the dominant classes in an advanced

---

8 "The walls that have separated the morro from the asphalt have always been invisible: walls of exclusion, of injustice, of omission and of terror. Now, the Government of Rio de Janeiro has decided to fully legitimate those invisible walls that already separated socially and spacely those two worlds of the city" (DE CARLI e HUMANES, s/d: 6).

9 The report launched in 2009 was organized by Rede Rio Criança, Projeto Rio Legal, PACS, IDDH, Lutarmada Hip-Hop, Visao da Favela Brasil, Mandato Marcelo Freixo, Observatório de Favelas, Associação pela Reforma Prisional, Justiça Global, ISER, Grupo Tortura Nunca Mais/RJ, APAFUNK, Movimento Direito para Quem - DPQ, MST/RJ, Centro de Assessoria Juridica e Popular Mariana Crioula, Central de Movimentos Populares - CMP, Rede de Comunidades e Movimentos Contra a Violência.

10 "This is not an exemplary wall, but a wall for containment purposes. Segregation and the expropriation of housing rights are effective. There is no need to try to guess what lies behind the supposed defense of the nature of ecolimits, or to try to infer what purposes are hidden in these good intentions" (CAMARGO, 2012: 2).

capitalist society, plays the role of legitimizing the state, which is also a representative of the elites, in the task of masking the true intentions of the ecolimites.

In this sense, the state, captured by the dominant economic classes, would have a single purpose and would form a cohesive whole in order to promote capitalist interests. Contradictions within the state are even admitted, but these would stem from its affiliation with the dominant classes[11] . The state is often treated as an entity that acts calculatingly in order to worsen the living conditions of the poor. When "differences" are perceived within the state, with the promotion of policies that, according to Melo (2010), bring different views of the favela, the state is considered contradictory .[12]

However, the state does not act univocally. Firstly, because in the case of the ecolimits, we are not dealing with just one government body, or even a single administrative level, but with a group of agents, politicians or secretariats, who have manifested themselves and produced meanings and practices relating to the implementation and design of the ecolimits in Rio's favelas. On the other hand, we don't think that the state has such a clear purpose that its actions can be reduced to a single objective.

As Pedroso also points out, the Legislative Assembly of Rio de Janeiro, through its Human Rights Commission and the mandate of state deputy Marcelo Freixo, organized, together with other human rights organizations, a report entitled "The Walls in the Favelas" which questions the construction of ecolimits and denounces their segregating nature. A public debate was also held by the Rio de Janeiro City Council

"(...) At the initiative of Councillor Eliomar Coelho (PSOL), the project to build ecolimits was deemed unconstitutional in this debate. Public defender Alexandre Fabiano Mendes, present at the event, explained that the construction of the walls violates the basic principles of citizenship, as well as going against aesthetic and cultural values" (PEDROSO, 2010: 26).

Considering these initiatives, it is possible to see the clash that takes place within what is called "the state". This demonstrates the conflict, the internal disagreements within the state itself, the possibility

---

11 "Within this analysis, the state is at the service of one class, the dominant class. Therefore, the state is full of contradictions, because although the state always serves the interests of its class, it has to claim that its actions are for the good of all. This causes a profound crisis of representation, since the dominated classes will not feel properly touched by the actions of the state. Therefore, the state has to adopt strategies so that people don't become aware of these contradictions and accept that social inequality, injustice and the truculence of the police in the favelas is something natural and necessary to contain the violence in these places" (PEDROSO, 2010: 54).

12 "Furthermore, the contradictory character of the official policies is not restricted to the State's duty of promoting societal well-being. It also refers to the apparently contrasting problem-framing that they unfold when looked at side by side. On the one hand, there are exclusionary policies of constructing walls surrounding the favelas and maintaining a special police unit that often violates human rights in its approach to the favela inhabitants; which disclose a perception of the latter, by the policy-makers, as enemies or outsiders, of troublemakers that do not pertain to the society and should hence be contained. On the other hand, there are also official policies whose objectives are to integrate the favelas into the city, to pacify them, to regularize and provide them with the infra-structure and public goods of a standard neighborhood. Examples of such policies are the Favela-Bairro program and the UPPs" (MELO,2010: 12).

of producing varied discourses that intersect and even oppose each other. Public policy not only acts as an instrument of persuasion, but can also incorporate elements brought about by the conflicts around it. In this sense, public policy does not evolve linearly, its end being determined by its initial objectives: the process of its construction can be marked by criticism, reassessment and ruptures.

3.3. The debate on ecolimits: towards a polyphonic object

Certainly, the issue of socio-spatial segregation plays an extremely important role in the debates organized around the announcement and construction of ecolimits. However, academia has disregarded the polyphony of politics and of public debate itself, producing a monochord discourse. Other spatial categories are called upon in this debate and make it possible to construct ecolimits as an object with multiple meanings and capable of raising other questions about the city. In a previous work mentioned above, we tried to identify some of the meanings attributed to the Santa Marta wall or ecolimit by reading the news on the *O Globo website*[13] . Here, we intend to point out the major axes of meaning of the spatial object in question. In other words, we will indicate the discursive axes recurrently used to deal with ecolimits in order to demonstrate the polyphonic nature of the object in question.

There are three types chosen because they constantly appear in the clash of meanings surrounding the wall in question. The first can be defined as the relationship between the city and nature or the forest. The second major type encompasses discourses dealing with the relationship between order and disorder in the urban space. The third and final axis focuses on debates about segregation, including those who believe that the wall is discriminatory and those who see no segregating intent in the construction of the ecolimit. These axes show that while the debate about segregation has become an issue in the city of Rio de Janeiro as a result of the announcement of the construction of new physical boundaries between favelas and green areas, it is not the only issue mobilized in the meaning of this project and the spatial categories it invokes.

The city and nature

The ecolimits project was promoted with the declared aim of containing the horizontal expansion of favelas over green areas and hillsides. We note that there is a relationship between the city and the forest: in this case, it is human presence that puts pressure on the forest, imposes danger, inspires care. This discourse is constantly mobilized to justify and make sense of the construction of the ecolimits. Even those voices that question the effectiveness of the walls point out the importance of preserving the Atlantic forest.

The ecolimits and the environmental discourse make explicit an inversion in the values related to the

---

13 The news was available at *www.oglobo.globo.com*

forest, to nature. It is no longer the unpredictable force of nature that must be contained and feared, but the impacts on green areas, especially on hillsides, that must be protected from human occupation[14]. In Rio, a city whose hallmark is this combination of city/built-up areas and green areas, the defense of the Atlantic forest gains legitimacy and is valued positively in various statements. However, the discourse of environmental preservation is not unanimous. Many voices contest this justification for the construction of the walls, such as the works cited above.

The order

In the news and articles about the ecolimits, it was possible to see the use of a type of discourse that has been generalized as the "discourse of order". This type of discourse is made up of statements that affirm the importance of order and regularization when dealing with the city and, more specifically, favelas. The use of verbs that indicate a positive value for public planning actions is quite evident, such as: prevent, inspect, combat, organize, urbanize, regularize, monitor, limit and legalize, almost always related to the state's action on favelas.

The discourse of order is not only linked to the construction of the ecolimit in Santa Marta. The various actions undertaken in this favela focus on order. This is the case of the Policia Pacificadora and the urbanization works of the state government, as well as the lighting projects of the municipal government. For this type of discourse, the favela needs to obey the laws, to function like the "asphalt". Above all, it needs to be contained, limited, so that it doesn't expand into the forest. In this way, the ecolimit acts doubly as the foundation of order because it clearly distinguishes (and creates) two spatial organizations - the forest and the city (even if the city is not yet legal, but is intended to be integrated into the formal city) - and limits the favela, regularizing and ordering its functioning and growth.

Segregation

Finally, we present the last axis: the discourse on socio-spatial segregation. The clash over this type of statement was widely publicized in the media, with characters from a wide range of fields taking part, such as José Saramago, a Portuguese writer, and Anthony Garotinho, a Rio de Janeiro politician, as well as representatives of the residents' associations involved in the construction of the ecolimits.

Some voices claim that the decision to build walls to limit the growth of favelas on green areas has a false justification. In other words, the environmental discourse is questioned as a way of "covering up" a policy that is not interested in dealing with the issue of poverty and inequality. Some articles resort to comparisons with historic walls, such as the one in Berlin. And many draw parallels with

14 While it is sometimes said that ecolimits also have the function of protecting surrounding residents from environmental risks such as landslides, this objective becomes secondary to the "need" to preserve the city's forest fragments.

ghettos, in order to defend the idea that ecolimits discriminate, segregate and deepen social and spatial differences. Some even question why walls are only built in the favelas. This is because approximately 70% of buildings above the *100 quota* are not classified as favelas.

It's interesting to note how the collected texts allude to the symbolic issue surrounding the construction of ecolimits. In other words, the dimension of meanings is also taken into account in the clash which is inevitably a political one, a confrontation of ideas, actions and projects for the city. Of course, other articles oppose these statements condemning the walls. These would be "politically correct", but would not adequately address the issue of preservation, which would be the real issue at stake.

3.4. Ecolimits: a discursive synthesis?

However, these discursive axes are not mutually exclusive. As we tried to demonstrate earlier, they can be integrated into a narrative that synthesizes some of Rio de Janeiro's central problems. None of these meanings is more true, but they are intertwined in the production of meanings for the city itself.

Based on the description of the material dimension of the wall, the actions associated with it and some of the meanings that were put into dispute through the news on the *O Globo* portal, we present a narrative that tries to respond to the intertwining of these three elements, the scenario[15] . A set of phases is proposed below, from a moment before the construction of the ecolimit to the prognosis of a later moment, produced by the images of the city, the wall, the favela and the green areas brought up by this public debate.

We start from the idea that there is a difference between the city and the forest. This difference is expressed when it is said that Rio de Janeiro is the result of a combination of a big city and the exuberance of the Atlantic forest. It's a kind of brand, an identity, a strong image for Rio. Therefore, there must be contact between these two distinct spatial orders (figure 3). In other words, the contact between the city and the forest has a positive meaning and is valued.

15 According to Gomes, "(...) with the word *scenario* we want to reconnect the physical dimension to actions, or, in other words, we want to associate spatial arrangements with behaviors and, from there, be able to interpret their possible meanings" (GOMES, 2008: 200).

Figure 3: the boundary between forest and city

However, there is another differentiation or boundary that makes up this scenario. It's the one that separates the formal city from the informal one, or the "asphalt" from the favela (figure 4). This boundary relates to different morphologies, actions, meanings and imaginaries relating to these two spatial orders.

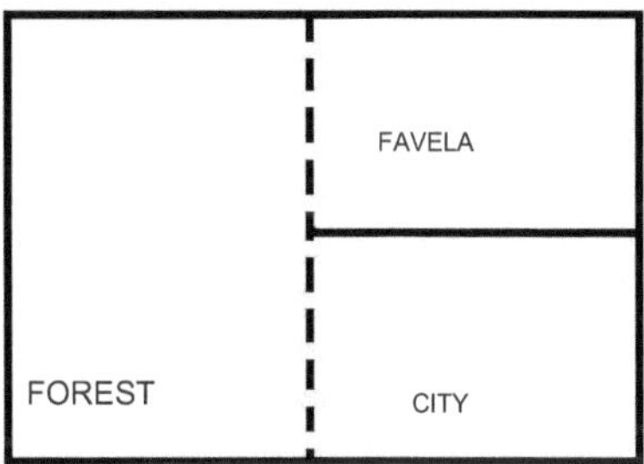

Figure 4: the forest-city boundary and the city-favela boundary

The actions already taken or deemed necessary in Santa Marta, Rocinha or other favelas are based on recognizing these differences in physical form, actions and values. Therefore, the actions must be different, the police must act differently, certain works must be done. The favela has another meaning, it doesn't have the same value as the formal city. It inspires fear, insecurity, repudiation; it can be considered a space of marginality, of illegality, but it is woven into the space of workers, of honesty, of the struggle against adversity.

According to some discourses, the favela expands into the forest, taking its disorder to new areas. It is in this sense that this space must be controlled, limited, prevented from growing, and the construction of the ecolimit is the strategy to solve this urban problem (figure 5). The ecolimit is a boundary between the favela and the forest, in other words, between a specific part of the city and the green areas. In this way, the difference between the favela and the formal city is present and reinforced.

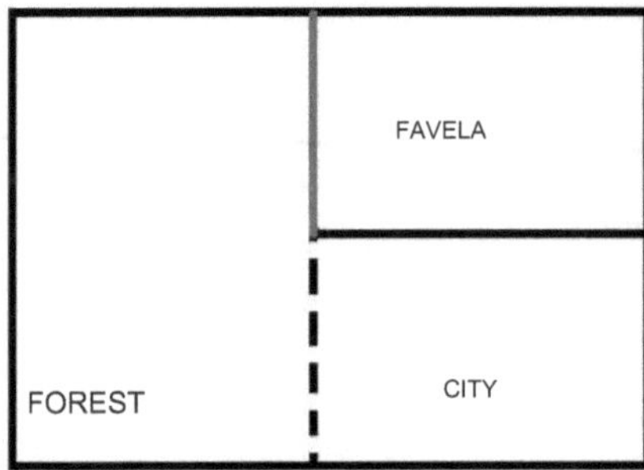

Figure 5: the ecolimit in the contact between the favela and the forest

However, the actions promoted by the public authorities appear to be intended to dilute this limit. The PAC works, the lighting and urbanization works, the entry of the police into Santa Marta and other favelas (including Rocinha, where the second ecolimit is under construction), the offer of wireless Internet in the favela: these initiatives seem to aim at a certain integration of this favela space into the formal city, based on the attenuation of the differences between their morphologies. If successful, these actions would hypothetically lead to the extinction of the second boundary (favela-city).

The construction of the wall is intended to preserve the forest. In a way, it aims to preserve the imbrication between city and forest that is so valued in Rio de Janeiro and, therefore, to preserve the city's own identity. The management of this interface and the ecolimit as an instrument for this management characterize a particular vision of urban growth, which turns towards the interior of the city, towards its internal discontinuities. Ultimately, ideally, the wall would be able to dissolve the boundary between the city and the forest because it would guarantee the image of Rio as this mixture, making the city a kind of harmonious and limitless unity (figure 6).

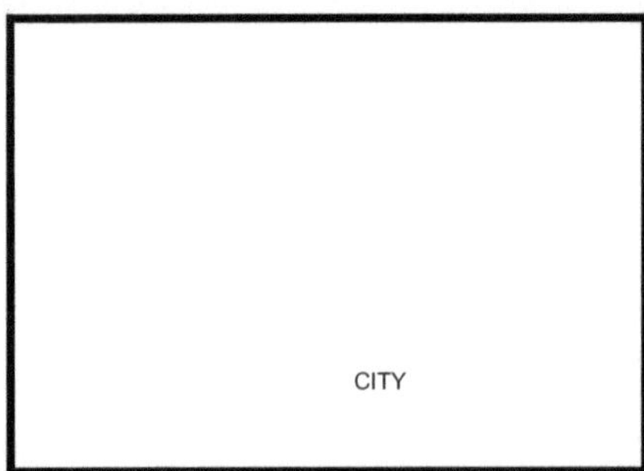

Figure 6: the dissolution of boundaries

In this way, the materialization and even the announcement of the implementation of the ecolimits are capable of putting into dispute various meanings between the various spatial categories mobilized in the construction of these limits that bring other ideas, other relationships that shape what we call the city and, particularly, Rio de Janeiro. The limits put existing spatial systems into relationship, requalifying them and the relationship itself. In other words, boundaries are capable of creating distinctions, of producing two or more spatial systems. In this sense, segregation is just one of the possible forms of the relationship between systems marked by boundaries.

An approach that takes into account the different spatial categories mobilized helps us to understand how they are constructed and how they are put into relation, what their physical and use dimensions are or should be and what their meanings and relationships are and, above all, what ideas of the city are put into play by the state in the different documents considered in the development of the research. This seems all the more interesting because we are dealing with boundaries, which place and create in juxtaposition differences of use, materiality and values that should be differentiated, opposed, but which are inevitably in relation.

Although the ecolimits have already been the subject of research in academic articles, monographs and dissertations, we believe that there are still possibilities for investigation offered by the conception and implementation of the project. If we take into account the discourses and practices produced about ecolimits by state agents, it is possible to see that they are capable of mobilizing issues of the utmost importance in the debate on urban policy in Rio de Janeiro: the growth of favelas and their relationship with the formal city, the management of conservation units in the urban space, preservation as a guarantee of environmental services, the appreciation of the relationship between built space and natural elements in the city of Rio de Janeiro.

4. Urban sprawl and containment policies

The containment of urban expansion for various reasons also seems to have been one of the motivations for the construction of physical boundaries. Even though they were not designed to limit the growth of cities, the walls had important effects on urban form. For Capel, "the importance of fortifications in the definition and evolution of urban form was great. In fact, the walls became anchor lines and barriers to the growth of the city."[16] (CAPEL, 2002: 137).

Roman cities, for example, were often planned to integrate unbuilt spaces that would allow for later expansion within the walled space. Even so, the creation of new suburban spaces outside the walls was frequent. Madrid, Barcelona, Florence and many other European cities, marked by demographic growth, built successive walls in order to include expansion outside the walls, the case of Florence being the most emblematic with the construction of seven walls. What was more common, however, was the existence of a Roman or high-medieval belt that would be extended once or twice during the Middle Ages.

"The number of defensive belts built by cities was logically affected by demographic evolution. Those that experienced sustained demographic growth were forced to build successive lines of fortifications. In Europe, numerous cities have built two, three, four and, exceptionally, even seven or eight defensive enclosures throughout their history" (CAPEL, 2002: 130).[17]

The walls could also be associated with preventing construction in large peri-urban areas. This type of restriction was due to the military interest in keeping this space around the city free of buildings that could be converted into protected points for enemy attack. In this case, urban growth was clearly linked to the need to defend cities. According to Capel, the walls became the anchoring lines of the cities and barriers to their own growth, even if they could be crossed or even moved. The use of these defensive belts was already practiced in the Neolithic period (cereal villages, as described by MUMFORD, 2008) and, in this sense, the author states that the city has had walls since its inception [18].

For Capel, in the 10th and 11th centuries many European cities could be classified as polynuclear and open. During the Lower Middle Ages, the many walls built surrounded these nuclei and created a unity based on the boundaries drawn by this closure. "From the point of view of the plan, this means

16 Free translation by the author from the original in Spanish: "(...) the importance of fortifications in the definition and evolution of urban form was great. In fact, the walls became a boundary line and a barrier to the growth of the city (...)".
17 Free translation by the author from the original in Spanish: "The number of defensive belts built by cities has logically been affected by demographic evolution. Those that experienced sustained demographic growth were forced to build successive lines of fortifications. In Europe, numerous cities have built two, three, four and, exceptionally, up to seven or eight defensive fences throughout their history".
18 "The technique of building defensive belts had already been developed and applied in the Neolithic period and in the age of metal to defend preurban settlements. And so the city has had walls since its very origins. They appear, in fact, in practically all archaeological excavations of primitive settlements" (CAPEL, 2002: 126).

the passage from the polynuclear or dispersed city to the compact city" (CAPEL, 2002: 129) .[19]

Many cities didn't have walls, either because they couldn't afford the huge investment that this type of construction entailed, or because they enjoyed a situation of natural protection or little hostility from neighboring peoples. On the other hand, the shape of these cities was also influenced by the absence of physical boundaries:

"(...) the absence of walls allowed for easier suburban expansion and the creation of the tradition of houses with gardens on the outskirts of the city. This is what happened in Britain, where, as has already been said, most towns did not have walls during the modern age" (CAPEL, 2002: 138) .[20]

In the bibliographical review, it was possible to find several works on the subject of containing or managing the growth of the modern, capitalist city. However, no works were found that analyzed, proposed or described the use of physical boundaries for these purposes, as in the case of ancient and medieval cities.

In short, the problem of containing urban growth already existed in ancient cities, in the Middle Ages and the Modern Age, and physical delimitation seems to have been the solution most widely put into practice. Many cities had their morphologies marked by the presence of walls or real wall systems that performed various functions and were associated with different meanings, such as protection from enemy peoples and wild animals, religious motivations (MUMFORD, 2008), collection of taxes and demarcation of areas with a governmental function, as in the case of Roman *citadels* or praetoriums, which were almost always associated with authoritarian regimes (CAPEL, 2002: 144).

Taking this into account, the interest here is to briefly present the debate not about urban growth, its causes and future scenarios, but about controlling the growth of the contemporary city through a series of instruments.

4.1. The terms of a debate

The current debate on extensive urban growth is generally structured from two perspectives. The first attributes a positive value to the spatial expansion of the city and, at the same time, questions the effectiveness of the instruments that seek to restrict it. The second defends the need to control urban growth, given the costs and problems it entails for urban life.

In the 1960s and 1970s, with the rise of the environmental movement and the emergence of the urban and economic crisis, concern about urban growth became more evident (BENGSTON et al., 2004).

---

19 Free translation by the author from the original in Spanish: "Desde el punto de vista del plano, significa el paso de la ciudad polinuclear o dispersa a la ciudad compacta."

20 Free translation by the author from the original in Spanish: "(...) the non-existence of walls could allow for easier suburban expansion and the creation of garden house traditions on the outskirts of the city. This is what happened in Gran Bretana where, as we said earlier, most towns did not have walls during the modern era."

Many governments (central and local) have responded to the issue with public policies aimed at managing extensive growth and protecting open spaces, in line with criticisms that identify a wide range of social and environmental costs associated with 'de-densified' cities. However, authors such as Glaeser and Kahn (2003) and Brueckner (2000) question the effectiveness and even the need for such policies, which result in a reduction in the availability of land, an increase in property prices and a reduction in their area.

According to Brueckner (2000), the conversion of agricultural land or natural environments into urban areas is not a disorderly process. On the contrary, it is regulated by the laws of the market, which direct resources to their best use. According to the author, there are three forces behind the growth of cities: population growth, rising household incomes and improved transport infrastructure. However, the market incurs three flaws when "calculating" the costs of urban expansion. The first can be described as a failure to consider the social value of open spaces that are converted to urban use. The second flaw stems from individual commuters underestimating the social costs involved in the intensive use of the road network. The third flaw is related to real estate developers' disregard for the public infrastructure costs involved in their projects. "In this sense, development seems artificially cheaper from the developer's point of view, which encourages excessive urban growth" (BRUECKNER, 2000: 163). For this author, the appropriate instruments should seek to correct the flaws arising from imperfect market assessment and should not focus directly on restricting the expansion of the city, as in urban containment policies.

Since the 1990s, a critical view of *urban sprawl* has emerged, based on the premise that extensive city growth results in the loss of the aesthetic benefits associated with the presence of open spaces, an increase in congestion and air pollution, and a reduction in arable land and incentives for the redevelopment of old centers. According to Bengston and Youn (2006), other negative impacts include higher infrastructure costs, increased travel distances, habitat loss, the risk of species extinction and damage to forest ecosystems.

With the emergence of the notion of *sustainable development*, criticism of urban sprawl has taken on new content, meanings and values. For Acselrad (1999), the encounter between sustainability and the problem of city growth stems from the search for legitimacy on the part of actors involved in the production of urban space. According to this perspective, the sprawling city would be incompatible with the ideals of the ecologically sustainable city (BARNIER and TUCOULET, 1999). "From a more technical point of view, the emphasis of the argument is on cost reduction and efficiency in the use of energy and transport resources. In this proposal, there is a clear assumption that a compact urban form produces greater sustainability" (COSTA, 1999: 65).

Urban form therefore plays a fundamental role in adapting cities to the terms of sustainable development. A sustainable city should be dense, especially around public transport hubs, and mixed uses should predominate, as well as offering access to recreational and sports areas, nature reserves and forests (COSTA, 1999; ROSALES and SANCHÉZ, 2011). In this context, instruments to contain growth, aimed at preserving rural areas and green areas and densifying underutilized areas, come to the fore.

4.2. How can growth be curbed?

For Bengston and Youn (2006), there are three common types of public policy instruments for managing urban growth. These are: public land acquisition, regulatory instruments and fiscal incentive instruments. The first type aims to protect open areas and can be carried out on land around the city or in areas inside the city. Regulatory instruments can take the form of controlling growth rates, prohibiting construction in certain areas, or zoning, land use planning and *urban containment* policies. Fiscal incentive instruments should encourage other development patterns and include property taxes and infrastructure impact fees.

For Pendall, Martin and Fulton (2002), any urban or metropolitan area has a policy to shape urban growth. Their objectives are diverse, but include preserving natural, agricultural or resource extraction areas, encouraging investment in already urbanized areas, creating a high-density land use pattern and stimulating mixed land use. The authors highlight three important instruments for shaping urban growth: greenbelts, *urban growth boundaries* and urban service areas. These instruments can be classified as *urban containment policies* and are often associated with other instruments such as zoning.

The green belt is the most restrictive of urban containment policies and originated in the 13th century BC in the cities of Palestine (BENGSTON and YOUN, 2006). It has been used in various contexts and can be defined as a strip of green space designed to protect open spaces or *working landscapes* around urban areas. *Urban growth boundaries*, on the other hand, are defined as a set of land use regulations that prevent urban development beyond a limit. They are usually associated with zoning. Criticism of this instrument refers to the difficulty of estimating the necessary or due expansion and, therefore, of locating the boundary. *Urban service areas* determine the limits up to which *urban* services (water, sewage and electricity systems) are provided. As they are more associated with the sequence of growth than with restriction, they tend to be more flexible and can be changed more easily (PENDALL *et al.*, 2002).

Urban growth management policies necessarily imply the creation of internal or external limits to the city, more or less flexible, material or not. This is made possible by the production of discontinuities between spatial categories that are also not given, but are summoned, created and transformed by the very production of public policy. In this sense, the notions of limits and discontinuity are central to the discussion of the Ecolimites project, as well as to the debate on urban growth management policies.

5. Elements for a conceptual discussion of ecolimits

This section is dedicated to a brief discussion of conceptual elements that enable the construction of ecolimits as research objects and help in their investigation. The first item proposes that boundaries, and therefore ecolimits, can be understood as instruments for spatial organization. The second discusses ecolimits as static discontinuities and producers of interfaces. The third and final section proposes that ecolimits should be characterized as devices, as proposed by Agamben.

5.1. The boundary as a principle of spatial order

The boundaries materialized in cities can take on different meanings. They can be used to demarcate the space of the sacred and of order, to protect against animals and invasions. They can be part of an ingenious defense system and organize relations with the space beyond the walls, the countryside, the forest (LE GOFF, 1998). They appear as components of the city, elements of cohesion and unification of the urban space and mark the border with the threatening exterior. They were and are used to organize contacts between specific groups and confer status on those who build them (CALDEIRA, 2000). They classify what should be inside and what should be outside.

Boundary strategies can also organize city growth, create obstacles and redirect occupation. Sometimes they are located at the outer limits of cities, sometimes they create boundaries within the city itself, producing internal differentiations in morphology, behavior and meaning.

The materialization of boundaries has an effect on spatial organization, creating and reconfiguring the relationships between the spatial systems in question. Boundaries and borders are elements of differentiation and operate as moments of divergence in the evolution of two territorial systems. Although these notions are strongly associated with the figure of the nation state, they play a part in the organization of space at other scales and in other situations, such as in the case of cities, or even in civilizations prior to the advent of the modern state.

They are often linked to ideas of segregation and interdiction of contact between groups and spatial systems, marked by images of walls that are very rich in meaning, such as the one in Berlin and those on the border between the United States and Mexico, and Israel and Palestine.

However, borders, limits and discontinuities create and reinforce distinctions between spatial organizations, delimiting differences between morphologies, behavioral aspects and meanings. They give meaning to the world, classify uses, groups and places, participating in its ordering. In discussing these concepts, other possible spatial relationships emerge, without the centrality of the sense of socio-spatial segregation.

According to Claude Raffestin (1992), the border is a territorial mediator that conditions systems of relationships. It should not be understood as an obstacle that limits individual and collective freedom, but as a regulatory mechanism that keeps us away from chaos and indistinction[21] . According to Reitel (2004), the border is a geographical object that separates two contiguous territorial systems, but which is not just the limit because it participates in the organization of space, integrating the political, symbolic and material dimensions. In this sense, the border can be understood as a discontinuity: that which separates two neighboring and different spatial sets (FRANÇOIS, 2004).

5.2. Boundaries, discontinuities and interfaces

According to Roger Brunet (1967), in his thesis entitled *"Les phénomènes de discontinuité en Géographie"*, discontinuities are intrinsic to social and natural phenomena. They are ruptures, leaps in the evolution of a system, causing it to change rhythm, meaning or nature. In concrete terms, discontinuities take the form of thresholds (*seuils*) that appear on a curve or in space .[22]

In an interview with François and Grasland for the magazine *L'Espace Géographique*, Roger Brunet makes a clear distinction between dynamic and static discontinuities. For the author,

"Dynamic discontinuities refer to processes: they mark a break in a movement. Static discontinuities are more or less durable local results of these processes. Static discontinuities are necessarily spatial: they are there, and there for a while (...). Therefore, their position can be the result of a whole dynamic, and we can build the hypothesis that these static discontinuities have their effects on systems, on dynamic discontinuities (BRUNET, FRANCOIS and GRASLAND, 1997: 301) .[23]

For the author, static discontinuities are part of differentiation systems and are capable of organizing space. They are not just barriers, but also places of passage and modes of continuity between systems. Discontinuities and thresholds represent ruptures within the same system, they are constitutive of the system itself.

One type of threshold described by Brunet is that of change of state[24] . The thresholds of change of state "(...) mark the limit from which a body or a phenomenon is transformed into another body or

21 "We must stop believing that the frontier is an obstacle and a counterintegration that opposes limitations to individual and collective freedom. The border is fundamentally a mechanism of regulation that guarantees existence against the dangers of chaos. La frontière est ubiquiste" (Raffestin, 1992: 163).

22 "(...) Silk is the concrete manifestation of discontinuity. Whether it appears on a street or in space, silk is the real mode of existence of discontinuity. Dans le texte de 1965, c'était solvente synonyme. Cependent, je préfère employer le terme de discontinuité dans son acception abstraite, pour évoquer um moment ou um point dans un processus, dans um mouvement d'ensemble; et le terme de seuil pour sa traduction concrète. Therefore, I don't like to talk about 'spatial discontinuity' for a specific geographical accident (except to refer to generality): rupture, trail, front, border, limit, border or contact, or even interface, are generally better suited and unambiguous" (BRUNET, FRANÇOIS and GRASLAND, 1997: 302).

23 Free translation by the author from the French original: "Dynamic discontinuities refer to processes: they mark a rupture in a movement. Les discontinuités statiques sont des résultats locaux plus ou moins durables de ces processus. Le discontinuités statiques sont necéssairement spatiales: eles sont là, et là pour quelque temps. (...) Cependent, leur position peut être le résultat de toute une dynamique; et l'on peut faire l'hypothèse que ces discontinuités statiques ont leurs efets sur les systèmes, sur les discontinuités dynamiques".

24When it comes to the relationship between thresholds and mechanisms, Brunet distinguishes three types: shear thresholds, change of state thresholds and relay thresholds.

another qualitatively distinct phenomenon" (BRUNET, 1967: 24)[25] . They mark a qualitative change and, therefore, the notion can be used to describe phenomena that are not measurable or that do not want to take into account quantitative aspects.

When dealing with concepts such as limits, boundaries and discontinuity, their separative nature is usually highlighted. However, as Ribeiro (2001) points out, solutions are produced for the continuity between two systems, which:

"They commonly emerge in the form of *intermediate* or *transitional zones* where phenomena of communication (more or less localized), diffusion, exchange, hybridization, etc. occur. It is for this reason that Roger Brunet (1992) suggests using the metaphor of the synapse to describe *border* phenomena, or even the concept of interface which, in the field of information technology, designates the connection between pieces of equipment that have different modes of operation" (RIBEIRO, 2001:7, emphasis in original).

According to Raffestin (1992), the border as an interface takes on the functions of translation, regulation, differentiation and relation. The function of translation refers to the boundary as a translator of information and intention. Regulation deals with the limit as a switch that opens and closes, allows or prohibits, guaranteeing the system's homeostasis. Assuming the function of differentiation against the chaos of indistinction, the border founds differences and preserves them. "Finally, the border is a relationship insofar as it juxtaposes territories that confront, compare and discover each other. Under these conditions, relations can be of opposition, exchange or collaboration" (RAFFESTIN, 1992: 160) .[26]

Ecolimits can be considered as static discontinuities or thresholds. This is because they separate contiguous territorial systems, reinforcing or even creating distinctions between what the morphological characteristics are or should be, the appropriate uses and behaviors in the preserved forest fragments and in the areas of urban occupation, thus producing distinct values. The ecolimits therefore mark a qualitative change within the system in question, the urban space in the city of Rio de Janeiro. They are localized products of this differentiation and, at the same time, act in spatial order.

The boundaries of Santa Marta and Rocinha function as qualifiers of the practices and other objects that make up the spatial systems they put into relation. The construction of the ecolimits organizes distinct spatial units, hierarchizes relationships and expresses certain values about the favela, the forest, the formal city and constructs many others. With the materialization of ecolimits, city and nature are not just abstract ideas, but acquire a morphology with limits, desired attitudes, specific and

25Free translation by the author from the French original: "Ils marquent la limite à partir de laquelle un corps ou un phénomène se transforment en un autre corps ou un autre phénomène qualitativement distincts."
26 Free translation by the author from the French original: "Finally, the frontier is relation in the sense that it juxtaposes territories that confront, compare and discover each other. Under these conditions, relations can be opposition, exchange or collaboration."

distinct values. These ideas and spaces are constructed in the relationship itself.

The walls built as part of the program act as territorial mediators, as a mechanism for regulating the relationship between "nature" and the "city". They constitute interfaces by producing, undoing and recreating connections between the two spatial systems that operate or start to operate in different ways. As well as regulating, relating and differentiating, they translate intentions, values and desires about the city. The ecolimits are not the interface itself, but what enables, regulates and manages it.

They are designed and built to regulate the relationship between green areas on hillsides and areas occupied by slums, a relationship that becomes a problem when we take into account the growth of slums over forest fragments. The conception of the project and the implementation of these boundaries respond to the desire, need or urgency of the interface management itself.

5.3. Ecolimits as devices

Considering the above, we propose that ecolimits be considered as devices, as defined by Agamben, one of the most important commentators on Foucault's thought today. For Agamben (2011), the term *device* becomes very important in Foucault's thinking from the mid-1970s onwards, when he begins to focus on questions of "governmentality". For Castro (2004), the *device* appears in Foucault's writings when he begins to focus more strongly on the analysis of power and the relationship between the discursive and the non-discursive. Or when his archaeology and, consequently, his object of archaeological description, the episteme, gave way to a genealogy, with the *device* being the object of genealogical description. Although he didn't define it clearly, he came close to the concept in an interview in 1977:

"What I am trying to identify by this name is [...] a deliberately heterogeneous set that includes discourses, institutions, architectural projects, regulations, laws, administrative measures, scientific statements, philosophical, moral and philanthropic propositions. In short, of the said and the unsaid, these are the elements of the device. The device itself is the network we establish between these elements [...]. By device, I mean a kind of formation which, at a given moment, had the predominant function of responding to an urgency. In this way, the device has a dominant strategic function..." (FOUCAULT apud AGAMBEN, 2007: 89) .[27]

According to Revel (2002), devices are as much about discourses as they are about practices, institutions and tactics. If we take ecolimits not as a single discourse or even as a single document, and consider that they are made up of newspaper reports, urban planning rules, environmental

---

27 Free translation by the author from the French original: "Ce que j'essaie de repérer sous ce nom c'est, [...] un ensemble résolument hétérogene comportant des discours, des institutions, des aménagements architecturaux, des lois, des mesures administratives.] a hétérogene résolument ensemble comprising discourses, institutions, architectural designs, regulatory decisions, laws, administrative measures, scientific concepts, philosophical, moral and philanthropic propositions; in short, from what is said as well as from what is not said, here are the elements of the dispositif. Le dispositif lui-même c'est le réseau qu'on établit entre ces éléments [...] par dispositif, j'entends une sorte - disons - de formations qui, à un momento donné, a eu pour fonction majeure de répondre à une urgence. Le dispositif a donc une fonction stratégique dominante..."

legislation, scientific research, moral judgments, architectural projects, we can say that ecolimits are constituted in this network of elements. These are discourses and practices that have different forms, produced by different subjects in different situations and aimed at multiple audiences. They are elements or discourses that interact, overlap, affect and cohere in order to respond to an urgency, to a question: the management of urban growth, in which one of the elements is the management and creation of interfaces, notably between environmental preservation areas and slums.

This management function assumed by ecolimits is a fundamental characteristic of devices. According to Agamben, when he describes the genealogy of terms such as *oikonomia*, *dispositio*, or the Heideggerian *gesell* (apparatus), the notions have in common a reference to the economy, management, government, control and orientation of men's behavior and thoughts. In Agamben's words:

"So, to give even greater generality to Foucault's *already* vast class of devices, I call a device everything that has, in one way or another, the capacity to capture, guide, determine, intercept, model, control and guarantee the gestures, conduct, opinions and discourses of living beings" (AGAMBEN, 2011: 258) .[28]

The device, although it produces subjects, is not based on being. It is that in which and through which the "activities of government" are carried out. They involve processes of subjectification, but they are the means by which this subjectification takes place, because they are not aimed at men, but at their behavior. Ecolimits, as a device, do not directly affect people or groups either. Their function is to regulate the contact between the two spatial systems placed in relation, organizing objects, behaviors and appropriate uses for each side, which certainly produces guidelines and constraints on men's lives.

The devices also have the characteristic of renewing and readjusting themselves according to the needs created by their own effects:

"(...) each positive or negative effect, desired or undesired, resonates or contradicts the others and requires readjustment. On the other hand, it is also possible to perceive a process of perpetual strategic fulfillment" (CASTRO, 2004: 148) .[29]

The device-ecolimites composed here is the network resulting from the disputes between the different statements considered, it is a mechanism for making and unmaking interfaces. Each of the elements that make it up and each stage of the project's implementation summon up and reformulate the terms of the debate or the two sides of the boundaries. Sometimes it dialogues with existing categories and

28 Free translation by the author from the original in Spanish: "So, in order to give greater generality to Foucault's vast class of devices, I call everything that has, in one way or another, the capacity to capture, guide, determine, intercept, model, control and ensure the gestures, behaviors, opinions and discourses of living beings a device".
29 Free translation by the author from the original in Spanish: "(...) each effect, positive or negative, wanted or unwanted, enters into resonance or contradiction with the others and requires readjustment. On the other hand, we also find ourselves with a process of perpetual strategic rellenamiento (remplissement)".

territorialities to create new boundaries and territories. At other times, it creates new spatial categories and reinforces established boundaries.

6. The device-ecolimits: elements and evolution

The time frame of the research was given by the set of elements that make up the ecolimits device. Thus, the device constituted here begins with the implementation of the first boundaries between favelas and green areas, the emergence of the term *ecolimites*, and its associated enunciations, in 2001. We consider this event to be a milestone in the production of public policies aimed at urban growth because it associates the management of the city's internal discontinuities with urban environmental problems, creating specific locations and materialities for the consolidation of boundaries. The end of the period in question was marked by the publication of the Sustainable Development Master Plan for the Municipality of Rio de Janeiro in 2011 (Complementary Law no.° 111, of 1°/02/11), the last published instrument to deal explicitly with ecolimits.

The spatial scope of the research was not established in advance of exploring the discourses considered as sources for the investigation. This is because we believe that this set of statements is capable of cutting out and putting into play different areas and territories, producing and undoing different spatial cut-outs. Thus, one of the questions pertinent to the subject is precisely the cut-outs, generalizations and spatial specifications produced by the device in question. In general terms, we can say that this section refers to the city of Rio de Janeiro and, more specifically, to Planning Area 2, where the ecolimits were installed.

The sources chosen for the research play a decisive role in the very configuration of the device described here. As they do not correspond to the totality of the discourses produced on the instruments aimed at containing growth or to the totality of the statements made by state agents, we will be dealing here with *a* device - ecolimits, a network of possible relationships between elements of a different nature. The choice made here cuts out the elements that directly concern the ecolimits, leaving out other elements that play a crucial role in the debate on the management of urban growth.

This set corresponds to the elements we were able to access through visits to municipal and state secretariats, bibliographic and newspaper research, laws and interviews. The absence or unavailability of official documents about the projects is noteworthy. As a way of obtaining information on the dimensions and conditions of the ecolimits' implementation, newspaper reports were used to constitute the device as elements which, despite their distinct nature, are important in the evolution of the policy. In this way, the device configured here included the documents that deal with the announcement and formulation of the projects implemented from 2001 and 2009 and the public policy instruments that function as antecedents and conditions for the implementation of the ecolimits.

The research tool designed aims to analyze the content of the documents/elements selected. It consists of a systematization sheet (which can be found in the annex) of the information collected according to the spatial categories mobilized, the boundaries presented and the interfaces produced. The tool also has a section for identifying the element itself, with fields for title, type of document, producing agent, date of publication, summary and other information. The categories were described in terms of their morphology, uses and related actions and meanings. The boundaries, in turn, were analyzed according to their functions, physical size, location, meanings and deployment conditions. Finally, the interfaces were described based on the related spatial categories, discontinuities and relationships between the spatial systems. The tool also contained a field for general comments about the documents or for information that did not fit into the previous sections. This instrument was applied to the selected documents.

The ecolimits device that will be described and analyzed in the following pages is a result of the projection of certain formulations onto the territory. Each time the physical delimiters, with their different morphologies and objectives, were projected onto the ground, there was a discontinuity in the evolution of the politics of ecolimits. The act of designing them, proposing them, building them, was an event that produced a rearrangement in the debate, in the forms, in the meanings of the ecolimits. With each announcement, project or wall built, a new moment developed.

The device-ecolimits is strongly marked by a temporal relationship that is established between the elements that make up a given moment. And each of these moments necessarily relates to other moments and, therefore, to other elements. If, on the one hand, the network that is formed between these elements can be described in chronological order, on the other, the moments intersect: they announce new trends, recover previous morphologies, mobilize similar categories, refer to meanings that have already been debated and criticize and incorporate each other.

Each of these moments, marked by elements capable of producing other discourses, mobilizes different categories and interfaces. By taking the moments as a whole into account, we believe it is possible to follow these movements between universalizations and specificities in terms of the categories, limits and interfaces used.

6.1. Eco-Limits: the 2001 program

According to officials from the Municipal Environment Secretariat (SMAC), the program drawn up by the Secretariat itself has not been consolidated in any official document that has been made public. The documents relating to the project are scattered and have not been organized. This can be seen by searching for information on the internet, on City Hall websites and in the database of urban legislation held by the Municipal Urban Planning Department, Busca Fàcil. Thus, the elements that make up this part of the device are eight newspaper articles that describe the formulation of Eco-Limits and present the debates that are mobilized by it. These articles were available on the O Globo newspaper's Agência O Globo database .[30]

The Eco-Limits proposal was conceived by the then Municipal Secretary for the Environment, Eduardo Paes. According to the first article that dealt explicitly with Eco-Limits, entitled "Favelas no limite: instalação de cercas de cabo de aço em 31 comunidades começa semana que próxima" (Favelas on the edge: installation of steel cable fences in 31 communities begins next week), on July 11, 2001, *Eco-Limits* correspond to *delimiters* and *fences* that have a defined materiality: they are a linear instrument made up of steel cables. In other elements, they have the morphology described by alambrados and rails with steel cables (O GLOBO, 26/05/2002), 630 meters of steel cables, in the case of Favela do Tirol (O GLOBO, 29/08/2002) and 600 meters in Portao Vermelho (locality of Rocinha) (O GLOBO, 02/10/2003b).

The implementation consisted of installing 24 kilometers of fences in *31 communities* in *22 areas of* the city of Rio de Janeiro, located *between the Pedra Branca and Tijuca massifs*. These areas are shown in a sketch (reproduced below - figure 5) and its title is "Learn more about the program" and its caption is "The areas covered by the project". This image is made up of four other elements: a box entitled "Where the occupation of the hills, above 100 meters, has been greatest in 15 years", a box entitled "Variation of vegetation cover in the city, in hectares", a text box on the fauna and flora of the hills of Rio[31] and another with data from the IPP Statistical Yearbook on the favelas[32] . The news item mentions *Rio das Pedras, Vidigal, Coréia* and *Piraquê favelas* and *small communities in Santa Teresa, such as Amapolo, Júlio Otoni and Mirante do Rato Molhado. Rocinha* is presented as the favela that would receive the pilot project and differs in that it has already had sections delimited in the past" (O GLOBO, 11/07/2001). In other elements, the ecolimits are located in *Areas of Environmental Interest*[33] , in *Rocinha*[34] and on the border between the community and the forest[35] , in the *South Zone*[36] , in *Favela do Tirol*[37] *, Vidigal*[38] *, Laboriaux* and *Portão Vermelho* (in Rocinha).[39]

30 All the articles available in the database containing the following words were searched: ecolimite, ecolimites, eco-limit, eco-limits.
31 *Rio's hills are areas of Atlantic forest. The municipality is going to reforest them with 36 native species, including sabià, maricà, angico, ipê and ingà. The fauna of the hills is basically made up of snakes, lizards, rats, skunks, marmosets, orange sabiàs, teiùs (a type of lizard) and boa constrictors.*
32 *According to the latest Statistical Yearbook, published by the Pereira Passos Institute, Rio has 462 slums, 256,586 households and a population of 952,429.*
33 THE GLOBE, 26/05/2002
34 O GLOBO, 26/05/2002 and O GLOBO, 04/08/2003
35 O GLOBO, 10/02/2003a
36 THE GLOBE, 26/05/2002
37 THE GLOBE, 29/08/2002
38 THE GLOBE, 04/08/2003
39 O GLOBO, 02/10/2003b

Saiba mais sobre o programa

Onde a ocupação dos morros, acima de 100 metros, foi maior em 15 anos

	1984	1999
Serra do Engenho Novo	5 ha (5,2%)*	17,2 ha (18%)
Serra da Misericórdia	8,4 ha (2,4%)*	40,7 ha (11,7%)
Morros isolados	5,5 ha (5,3%)*	12,3 ha (11,9%)
Maciço da Pedra Branca	29,3 ha (0,7%)*	115,5 ha (0,9%)
Maciço da Tijuca	507,8 ha (5,4%)*	251,7 ha (8%)

* da área total

Variação da cobertura vegetal na cidade, em hectares

Tipo de vegetação	1984	1999
Floresta	30.300	25.243
Mangue	3.552	3.489
Restinga	1.103	771
	Total: 34.955	Total: 29.502

Perda: 5.453 (15,6%)

Os morros do Rio são áreas de mata atlântica. A prefeitura vai reflorestá-los com 36 espécies nativas, entre elas sabiá, maricá, angico, ipê e ingá.
A fauna dos morros é formada basicamente por cobras, lagartos, ratos, gambás, micos, sabiás-laranjeira, teiús (um tipo de lagarto) e jibóias.

Segundo o último Anuário Estatístico, do Instituto Pereira Passos, o Rio tem 462 favelas, com 256.586 domicílios e população de 952.429 habitantes

1. **Rio das Pedras**: 200 metros de trilho com cabo de aço
2. **Estrada do Sertão**: mil metros de trilho com cabo de aço
3. **Sítio do Pai João**: 600 metros de trilho com cabo de aço
4. **Rocinha**: 60 metros de alambrado
5. **Corredor ecológico**: 15 mil metros de trilho com cabo de aço, marco e alambrado
6. **Rua Tirol**: mil metros de trilho com cabo de aço
7. **Estrada do Tijuaçu**: 600 metros de trilho com cabo de aço
8. **Muzema**: 600 metros de trilho com cabo de aço
9. **Coréia**: 1.020 metros de trilho com cabo de aço
10. **Piraquê**: 2.055 metros de montantes de concreto resistente a salitre
11. **Sumaré**: 210 metros de trilho com cabo de aço
12. **Vidigal**: 800 metros de trilho com cabo de aço
13. **Jardim do Carmo**: 1.400 metros de trilho com cabo de aço
14. **Rua Santa Alexandrina**: 565 metros de trilho com cabo de aço e alambrado
15. **Mirante do Rato Molhado**: 230 metros de alambrado
16. **Estrada Joaquim Mamede**: 1.365 metros de de trilho com cabo de aço e alambrado
17. **Rua Professor Olinto de Melo**: 850 metros de trilho com cabo de aço e alambrado
18. **Rua Sinimbu**: 300 metros de trilho com cabo de aço
19. **Ricardinho**: 600 metros de trilho com cabo de aço
20. **Júlio Otoni**: 220 metros de alambrado
21. **Vila Alice**: 600 metros de trilho com cabo de aço
22. **Condomínio Village 400**: 30 metros de alambrado

Figure 7: Infographic presented in the first Eco-Limits news report

Source: O GLOBO, 11/07/2001

The aim of installing these *delimiters* was to *"stop the expansion of favelas"* (O GLOBO, 11/07/2001) and *to isolate favelas in order to contain the expansion of communities* (O GLOBO, 04/08/2003). In other elements that make up this moment of the device, the objective of the ecolimits is oriented towards environmental preservation and the physical delimitation of green areas. In this way, the ecolimit would aim to isolate green areas (O GLOBO, 01/06/2003), physically delimit green areas (O GLOBO, 29/08/2002) *and maintain environmental protection areas* (O GLOBO, 02/10/2003).

Sometimes, the stated function of ecolimits relates categories such as *slums* and *green areas*, describing the interface between these types of spatial categories. These formulations can be presented by the following examples: *fencing off areas of forest that are being invaded by favelas, helping to preserve what remains of the forest and at the same time preventing the uncontrolled expansion of irregular housing* (O GLOBO, 26/05/2002) and *"preventing favelas from continuing to expand over green areas"* (O GLOBO, 11/07/2001).

Considering the name given to the project by the Municipal Secretary for the Environment in 2001, Eduardo Paes, who associates the prefix *eco* with the limits, it is possible to say that it has a value associated with the ecological issue in urban environments. Thus, the project would carry with it a very positive valuation, given the importance given to environmental problems in cities since the 1960s and more strongly since the 1990s. The first report also includes the opinion of the president of the Rio das Pedras Residents' Association, one of the favelas that would be targeted by the project. According to the report, "Josinaldo Francisco da Cruz, or Nadinho, *applauds* the Eco-Limits program". By presenting this favorable opinion from a resident and representative of the favelas, the delimiters gain acceptance and an apparent unquestionability. In 2002, according to the then Municipal Secretary for the Environment, Pedro Paulo Teixeira, this is a cheap project if you consider its benefits and scope (O GLOBO, 26/05/2002).

Despite these opinions in favor of the project and the meanings that legitimize it socially, the ecolimits have also been criticized since 2003. These criticisms are related to the project's ineffectiveness in combating attacks on environmental preservation areas[40] and in containing the growth of favelas. This is because the delimiters implemented within the framework of the program are being ignored (with houses being built beyond what is allowed in Rocinha, for example)[41] and are only aimed at containing the horizontal growth of the favelas, neglecting what has been called the vertical growth or verticalization that is taking place in these areas .[42]

At this point in the device, the spatial categories mobilized can be divided into two large groups. One is made up of categories linked to slums and irregular occupations and the other to areas with environmental value. The first group includes *irregular occupations*, *favelas*, *Rocinha*, *irregular housing*, *registered favelas*, *informal occupation area*, *Vila Alice*, *Vidigal*, *Favela do Tirol*, *Favelas in the* (Barra) *region*, *areas occupied by favelas*, *Laboriaux in Rocinha*, *Morro do Banco*, *Rio das Pedras*, *other smaller favelas*, *shanty towns*, *Vila Autódromo*, *Portão Vermelho*, *communities*, *undue occupations*, *deprived communities* and *new communities*.

In general, these categories are characterized by a morphology of uncontrolled expansion, which often takes place over green areas, as they are located on hillsides, forests and mangroves, in or near areas of nature conservation units. Another characteristic is that some of these favelas, as they grow, could join others. These are places where eco-limits have been or will be set up (in the case of Rocinha, as a pilot project), which cause deforestation and disrespect urban planning rules. They are seen as invaders and must be combated. They pose a risk to areas of environmental preservation and therefore their growth is worrying. They are associated with ideas of disorder and pollution.

The second group, green areas, was made up of the following categories: *green areas*, *forest*, *Tijuca National Park*, *lagoons* (Tijuca and Jacarepaguà), *areas of environmental interest*, *remaining forest* and *environmental protection areas*. Their morphology is characterized by occupation or invasion by favelas, by the need for demarcation or fencing. These categories are also qualified by the reduction

---

40 This and other projects have not yet managed to overcome the attacks, according to an article in O Globo on June 1, 2003.

41 O GLOBO, 02/10/2003b

42 *The main criticism of Eco-Limites is that it doesn't prevent the vertical growth of favelas* (O GLOBO, 04/08/2003).

of their areas, deforestation and the dumping of garbage.

These are areas where there should be no illegal or irregular construction. The devastation that takes place there is a point of divergence: sometimes it is *visible*, sometimes it has been *stopped*. These categories are also marked by degradation and must be protected from the disorderly growth caused by irregular occupations. In the case of the Rio das Pedras favela, *solving the environmental problems of Barra's lagoons will help with the bid for the 2012 Olympics* (O GLOBO, 04/08/2003), which makes clear the relationship between the environmental issue and a positive vision of the city, which must be competitive on a global scale.

There is also the mobilization of categories that can be considered mixed, or interface categories. Not because they mean the synthesis or overlapping of favelas and green areas, but because they reveal the relationships between these spatial categories. These are the *green areas that have not yet been invaded*, the *favelas next to the city's green areas* and the *forest areas that are being invaded by favelas*.

Another spatial category present at this time is the *ecological corridor*. Located between the Tijuca and Pedra Branca massifs, its morphology would be defined, with the implementation of the project, by 15 km of *trails with steel cables, fencing and landmarks*. This would lead to the eviction of 30 houses in the area. This category is also mobilized in the article entitled "Until September, only 5% of favelas surrounded", which presents the neighbourhoods through which the corridor would extend, as well as reporting on the occupation of the area by *ten poor communities* (O GLOBO, 26/05/2002).

It is also possible to highlight the category of *luxury homes*, which cannot be included in the first group of spatial categories, although it does have some of their characteristics. The category is mobilized by the article entitled "One fifth less green areas: irregular occupations invade protected areas". These luxury homes are located in Sao Conrado and, according to the caption of the photo illustrating the article, they *were built on the slopes of Pedra da Gàvea, above the 100 level and within the protected area of the Tijuca Forest* (O GLOBO, 02/10/2003). Like the favelas, they are characterized by a lack of respect for urban planning rules .[43]

Considering the large groups of spatial categories defined above, the privileged interface is between *green areas* and *slums*, and its discontinuity is the eco-boundary. The relationship between these spatial systems is that of the continuous expansion of the latter over the former, with the growth of irregular occupations having to be contained by the delimiters known as eco-boundaries. Slums, irregular and undue occupation put preservation areas, the Tijuca National Park and remaining forest

---

43 In this case, the disrespect is in relation to the limit established by the 100-meter quota and not by the installation of ecolimits.

areas at risk. This creates a scenario of *urban disorder* (O GLOBO, 02/10/2003a).

Other more specific interfaces are also produced at this time. This is the case between *Rocinha* and the *forest*, which also has an eco-limit as its discontinuity. The relationship between these categories is that of the advance of the shacks over the forest, ignoring the installed landmarks (O GLOBO, 02/10/2003a). In this and other news reports, there are clearly other interfaces, such as the one between the *Tijuca Forest* and *poor communities*. However, the boundary between these two spatial categories is not explicit. It is only noticeable that these categories differ, relate and are juxtaposed. In this case, the relationship is marked by the invasion *of poor communities* into the *Tijuca Forest* area.

Another specific interface that can be highlighted is the *Rio das Pedras Favela* and the Barra *Lagoons*. In this case, the boundary itself was adapted according to the categories mobilized: according to the then president of Serla, Icaro Moreno Júnior, instead of steel cables, the Rio das Pedras ecolimits would be *made of 300 eucalyptus logs* at 100-metre intervals (O GLOBO, 04/08/2003). The interface can be defined as follows: the *Rio das Pedras Favela* and *other smaller favelas threaten the* water mirror and *degrade the* environment of the lagoons.

Considering the elements of this moment of the device described above, the agents that stand out are related to environmental management entities, the Municipal Department of the Environment and the State Foundation for Rivers and Lagoons - SERLA, in the form of secretaries Eduardo Paes and Pedro Paulo Teixeira and president Icaro Moreno Júnior. In addition to these, we could highlight the favela residents' associations, represented by the president of the Rio das Pedras Residents' Association who, in one of the articles, was invited to give his opinion on the project.

This period sees the mobilization of general and specific spatial categories and interfaces. In the first news item, when the programme is presented, the general categories and interfaces that organize the intervention are presented. Later, with the implementation of the ecolimits on the ground, the categories become more specific, bringing up issues that are particular to each situation. This is the case of *Rio das Pedras*, where the morphology of the project is adapted in view of the spatial category *Lagoas da Barra,* which differs from green areas or forests, resulting in the production of a particular interface. *Rocinha* is another example of a specific category, notable for being the favela that would receive the pilot project, for having already received delimitations that would indicate the antecedents of Eco-Limits and for also being described in terms of its localities, such as *Laboriaux* and *Portão Vermelho*. Criticism of the program also emerges from the evaluation of specific situations, which produce interfaces of the same character.

Based on the identification of the problems and values associated with the program, it is possible to make some considerations about the vision of the city that would be at stake. On the one hand, the expansion of favelas is seen as a problematic situation that must be solved by physically delimiting their areas. In this way, the causes of the emergence and growth of slums are not addressed and containment is presented as a measure that indicates the values that make up the spatial categories related to slums: these are threats to the urban environment that must at least be contained if not combated. On the other hand, the Eco-Limites program is strongly associated with the environmental issue, claiming the need to preserve green areas, forests and water bodies for the benefit of the city and its image in the world.

6.2. A wall for Rocinha

In 2004, the issue of ecolimits once again came to the fore in the public debate with the announcement of the construction of walls to contain the growth of the Rocinha favela. As in the previous period, no official document dealing with the project has been identified. The proposal, however, was reported in the newspaper O Globo, as were the discussions arising from this intention. Thus, the element that triggers this moment of the device is the announcement of the construction of the walls, which will be taken up again through the newspaper reports. These, in turn, are part of the device constituted in this dissertation.

The proposal to build a wall to surround Rocinha and three other favelas was published by the newspaper O Globo in a report entitled "A Guerra do Rio: limite da violência" (Rio's war: the edge of violence), on April 12, 2004. The project was authored by the then Deputy Governor and State Secretary for the Environment, Luiz Paulo Conde, and was to be located in Rocinha and three other favelas: Vidigal, Chácara do Céu and Parque da Cidade. The walls that would surround parts of the favelas should be 3 meters high. According to the news report, in Rocinha, *a path will be built along the wall for cars to pass through, two and a half meters wide* (O GLOBO, 12/04/2002).

The objectives of the project are only described for Rocinha and are linked to the growth of the favela over the Atlantic Forest area and the actions of drug traffickers. At the time, Rocinha was suffering *from a war for control of the drug trade* and the construction of this wall was a response by the government to the situation. The caption of the photo that illustrates *the* article sums up this double objective: *in Rocinha, the expansion of houses into the Atlantic Forest area would be prevented by the construction of a three-meter high wall, which would also serve to prevent bandits from hiding in the middle of the bush.* According to the subtitle, this wall is a *limit to violence.*

For Conde, it was *a measure to protect the residents of these communities.* The project would be able to bring *peace of mind* to residents and police officers who walked along the wall. Thus, the walls would benefit the residents of the favelas and would be approved by the population: *the delimitation of these favelas would benefit their residents and the population, in general, would applaud it.*

Other agents are also interviewed in this report and give their opinions on the project. According to architect Sèrgio Magalhaes, from the State Secretariat for Urban Development, *the idea is perfectly feasible* and it was necessary to contain the expansion of the favelas into the forest. The president of the Rocinha Residents' Association, William de Oliveira, was in favour, although he recognized the need for the *community* to be consulted.

Voices against the implementation of the walls are also heard in this first report. For Alessandro Molon, then a state deputy for the Workers' Party (PT) and president of the Legislative Assembly's Human Rights Commission, the idea represented *a testament to the authorities' unpreparedness to deal with violence.* For Leonarda Musumeci, a professor at UFRJ and a researcher at the Centre for Security and Citizenship Studies at Càndido Mendes University, the wall is *reminiscent of separation, apatheid* and symbolizes the *bankruptcy of the state.* For Eduardo Paes, who conceived the Eco-Limits project in 2001 as municipal secretary for the environment and at the time was a federal deputy, the idea was *ridiculous*: *a wall wouldn't stop bandits from crossing from one side to the other, no matter how high it was.* Airton Xerez, then Municipal Secretary for the Environment, who continued the Eco-Limits project, declared: *the government seems to be adopting the extermination policy of Ariel Sharon and Hitler.*

At this time, as well as when the Santa Marta ecolimit was built (later, in 2009), the walls built in various parts of the world with clear intentions of separation or segregation are brought into the discussion. This news item features an explanatory box entitled *Barriers in the world*, which mentions the walls of Berlin, Warsaw and the West Bank and begins with the following sentence: *In the world, walls are built in the name of intolerance.*

The article presents the Eco-limites program as an antecedent of this project or something comparable to it with the intertitle *Prefeitura criou programa parecido.* This relationship can also be seen in the presentation of the opinions of agents deeply involved in the conception and installation of the eco-limits (Eduardo Paes and Airton Xerez), which are described as *delimiters* intended to prevent favelas from continuing to expand over green areas. Their morphology was described by iron rails interconnected by steel cables. Since 2001, around 30 kilometers had been installed *in more than 40 communities* and, at the time, another 11 kilometers were expected to be installed. According to the municipal environment secretary, Airton Xerez, the fences were being respected in most communities. The report also highlights the fact that *the first fences were installed in Rocinha.*

In this element of the device, the following spatial categories are mobilized: *Rocinha, Vidigal, Châcara do Céu, Parque da Cidade, favelas, communities, environmental protection area, forest* and *Atlantic Forest area. Rocinha* is characterized by expanding into the *Atlantic Forest area* and part of the *hill* would be surrounded by walls 3 meters high. It destroys an *area of environmental protection* and is presented as a violent space, which has repercussions not only in Rio, but throughout the country. According to Conde, *"The violence in Rocinha is not a problem specific to Rio. It's a problem that affects other states. It's a national issue"*.

Vidigal, *Chácara do Céu* and *Parque da Cidade* are presented as *three other favelas* classified as destroying an *area of environmental protection* and should be partially walled off. These are *favelas* and *communities* that are expanding into the *forest* and must be contained if the infrastructure services are to be successful.

The *Atlantic Forest* Area is occupied by houses from Rocinha. The *Environmental Protection Area* category is characterized by the destruction it suffers from the four favelas mentioned above. The area of the *forest* is occupied by the expansion of the favelas and is used as a path for incursions by drug traffickers. The interface present in this element of the device is marked by the *expansion, occupation and destruction of* the *favelas/communities Rocinha, Vidigal, Chácara do Céu and Parque da Cidade* over *areas of Atlantic Forest/environmental protection* that should be preserved by the construction of the walls. However, this interface can only be understood if we take into account the meanings and uses (violence, illegal activities, presence of drug dealers) attributed to these two sets of categories.

Two days after Conde's proposal was made public, on April 14, 2004, a short article of just three paragraphs was published, reporting on the appearance of yet another house in the middle of the forest in Rocinha. The last paragraph deals with the Eco-Limits Program and its installation in the same favela. Rocinha, the *first favela in Rio where the city government implemented the Eco-Limits program,* was moving beyond the landmarks. These delimited the areas beyond which the favelas should not expand and were made up of *iron stakes connected by cables.*

The following day, in a story entitled *A Guerra do Rio: technicians tried to inspect property but left when the climate became tense in the favela,* the construction of houses beyond the limits in Rocinha is again reported. In the favela, there were 65 buildings erected *after the Eco-Limits,* an area into which the favela could not expand. In this area, a house is being built that could house drug traffickers. The article also presents the localities *of Dionéia*, used as an escape route for bandits, and *Laboriaux*, accessible from Rua Dionéia through the woods and where the head of the favela's drug trade had been killed the day before. It is interesting to note, albeit inconsistently, that the forest is not necessarily opposed to the favela (mobilization of the spatial category *forest over favela*). In this sense, the houses are built or the favela expands beyond the limits defined for its growth, but into an area that, by a play on words, is also part of its space.

For Sirkis, then Municipal Secretary for Urban Planning, who signed the article on May 3, 2004, *the horizontal growth in this favela has been contained since 2001, with the implementation of the ecolimits*. For the secretary, Rocinha has progressed in terms of infrastructure and income-generating projects, has experienced a rise in the standard of consumption of its residents and has different values from those cherished on the asphalt. In the text, ecolimites are understood as *borders*. In this way, Rocinha doesn't expand into *the green areas* or *outside this border*, except for the 63 houses that are beyond the ecolimits. The favela would be controlled in its growth, no longer occupying undue areas.

In July 2004, a news report about the APARU do Alto da Boa Vista was published with the title *Projeto Ecolimites detains favelas: Aparu do Alto, still being debated, plans to build on hillsides*. Ecolimits are described as *artificial barriers, which can be made of steel cables, fences or concrete markers*, with the aim of combating slumification or stopping slums, according to the title of the article. They are set up in *areas where slums are expanding*, as in the case of APARU in Alto da Boa Vista. In this area, located on *hillsides* characterized by the presence of native vegetation, slum occupation was *being combated with the Ecolimites Project*. The interface present there can be described as the growth of slums on the hillside areas. However, in this situation, the ecolimits are valued for their efficiency in combating this growth (O GLOBO, 01/07/2004).

In 2005, ecolimits continued to be debated and the Rocinha favela featured prominently in the news in the newspaper O Globo. This year, between September and October, the construction of an eleven-storey building in the favela became known as the "Rocinha spire" or "Rocinha Empire State". This resulted in a debate about the illegality and irregularity that characterized the favela, and its need for urban regularization.

This year, the news reports that relate the project to Rocinha describe the Eco-Limits as fences *that isolate parts of Rocinha occupied by houses from the green areas* and are intended to *prevent horizontal growth*. However, as the supervision by the technicians of the Municipal Urban Planning Department is concentrated on these limits, it does not prevent vertical growth (O GLOBO, 23/09/2005a). Its function is also described, in the case of Rocinha, by the demarcation of areas where houses must be removed (O GLOBO, 13/10/2005).

In another article, the Eco-Limits are described as an *illusory success*, according to the headline. It goes on to criticize the effectiveness of the ecolimits in containing the vertical growth of the favelas: *the city government says it is a success. It would be, if it also contained the growth of favelas upwards* (O GLOBO, 23/09/2005b). On September 27, 2005, Alfredo Sirkis, then Municipal Secretary for Urban Planning, defended the efficiency of the *Eco Limits* in Rocinha: *the comparison of satellite photos from 1999 and 2004 proves that the horizontal expansion was contained by the Eco Limits, which, until recently, many thought impossible.*

In the 2005 news, as mentioned above, the spatial category *Rocinha* is mobilized significantly in the debate about ecolimits. Its morphology is marked by vertical expansion and sprawl *across the entire hillside,* according to the caption of the photo illustrating the September 23 story (O GLOBO, 23/09/2005a). *The slabs are being stacked indiscriminately*, highlighting their vertical growth (O GLOBO, 23/09/2005b).

Rocinha, where there are no rules and control by the government (O GLOBO, 23/09/2005b), has shacks built *outside the ecolimits.* The removal, which was ordered by FECAM, would help to *recover the Atlantic Forest* (O GLOBO, 13/10/2005). These houses were located close to the boundaries of Tijuca Park (O GLOBO, 16/10/2005), in an area known as *Laboriaux*, at the top of the favela. By disrespecting the boundaries, they turn the area "outside the ecolimits" into an area of the favela itself (O GLOBO, 13/10/2005).

In September of the same year, *the state government and the Institute of Architects of Brazil* launched a *design competition with proposals to create an urbanization plan for Rocinha, a competition to change the community* (O GLOBO, 23/09/2005a). The winning firm was to produce a plan with proposals for favela urbanization that could be adopted to guide the construction work.

Still on the subject of *Rocinha*, it's possible to notice another type of discourse related to its merger with other favelas. *Vidigal, Rocinha, Vila Parque da Cidade and Chácara do Céu* were advancing horizontally and could form a complex within ten years. The merger between *Rocinha* and *Parque da Cidade* would take place if an *area of forest* was invaded, which meant a barrier to the merger of these favelas. The formation of this complex would damage the population living in the favelas and the image of the city: *the safety of the population living and building in risky areas and the beauty of Rio are at stake* (O GLOBO, 09/10/2005a).

For Marina Magessi, then an inspector with the Civil Police, the formation of this complex would have to involve a reconfiguration of the power of the drug trade. She regrets that the American School has left Rocinha, as it acted as a boundary *to the total înlegracăo with the City Park.* Magessi describes the eco-limits as a *joke.* For the inspector, *everyone knows that the favelas are moving forward and will meet* (O GLOBO, 09/10/2005b).

This concern about the merging of favelas is also associated with other spatial categories. *Vila Alice* and *Jùlio Otoni* are qualified by the concern about their possible merger (O GLOBO, 07/10/2005). In another report, the issue of merger reappears in relation to the category of *favelas in Santa Teresa,* which showed uncontrolled expansion that threatened to *unite the local favelas with those in neighboring districts* (O GLOBO, 26/10/2005).

Vila Alice, which is located in the *São José Environmental Preservation Area, an area in Laranjeiras where eco-limits are not respected,* is also characterized by growth that *borders a neighboring community, Jùlio Otoni.* The favela area is marked by a fierce dispute between *the owners of the land,* who are trying to *get it back in court*, and its residents, who reject the removal proposed by the City Hall.

Its ecolimits had been implanted in 2002 and *lost their purpose,* gaining *new uses (...): support for clotheslines and sinks, as well as a foundation for a shack* (O GLOBO, 07/10/2005).

In the report on the *favelas of Santa Teresa, Pau da Bandeira* and *Pereira da Silva* are merging, and the boundary between *Morro dos Prazeres* and *Escondidinho is difficult to distinguish. Júlio Otoni*, on the other hand, goes through the *São José APA* and approaches *Vila Alice.* In the *Coroado favela,* where the Eco-limit was implemented and benefited the green area around the community, the project was *unable to contain growth* (intertitulo), as the favela kept growing vertically. According to Flàvio Minervino, president of the Santa Teresa favela coalition, *the communities are growing because the city government can't control the expansion* (O GLOBO, 26/10/2005).

Morro Azul, in the Flamengo neighborhood, is described as a *degraded area* and a *poor community*, where the Bairrinho program was being implemented. The favela was to be integrated into the *asphalt* and was growing at a rapid pace. This was because the ecolimits set up there were unable to control vertical growth (O GLOBO, 28/10/2005).

Also in 2005, articles were published that dealt with the ecolimits in a more general way, producing evaluations of the project. This is the case of *Eco-Limites has reached 50 communities*, an article that describes the ecolimits with the aim of *containing the expansion of favelas into green areas.* The project, which could be expanded, had already installed *more than 40,000 meters of steel cables and fences* in 50 *poor communities in Rio, such as* Rocinha, the Itanhangà and Jacarepaguà favelas, Vila Alice, Júlio Otoni, Rato Molhado, Estrada do Sumaré and *30 other threatened green areas*. The interface here is between *favelas/poor communities in Rio and green areas/threatened green areas.* The favelas threaten the green areas through their expansion and, in this sense, must be contained (O GLOBO, 07/10/2005b).

The article entitled *Illegal. So what?: favela expansion has no eco-limits. An audit by the Court of Auditors proves that communities are rapidly advancing into preservation areas,* mobilizing more specific categories such as *Laboriaux* and *Babilònia*, but also more general categories such as *17 favelas* and *42 communities.* The *17 favelas* occupy areas of environmental preservation and are marked by expansion that *has no ecolimits* (subtitle) and many of which put a *rich ecosystem at risk.* The *42 communities* category is qualified by the situation at a maximum distance of one hundred meters from areas administered by the Union, the state or the municipality. The *environmental preservation areas* category *is* described as areas that are at risk and must be preserved. This relationship between *favelas/communities* and *environmental protection areas* suggests the emergence of a new category, that of *favela reserves.* For the then president of the Legislative

Assembly's Environment Committee, Carlos Minc, *we run the risk of these parks being transformed into what I call slum reserves, due to a lack of control over expansion* (O GLOBO, 16/10/2005).

In 2006, the news about ecolimits was mainly related to the *Rocinha, Laboriaux, Tijuca Forest, Tijuca Massif Forest* and *Alto da Boa Vista APARU* categories. *Rocinha* is growing in a disorderly fashion towards the asphalt and the protected areas. The residents reject this type of growth and their *main demand (...) is for the development of their own urban legislation to be discussed with the community* (O GLOBO, 09/02/2006). The *Laboriaux* area is characterized by uncontrolled growth towards the Tijuca Massif (O GLOBO, 10/08/2006). At that time, the favela would receive a major urbanization project and, for architect Luiz Carlos Toledo, coordinator of the team that won the competition promoted by the IAB and the State Government the previous year, it *is not a separate universe from the city* (O GLOBO, 31/10/2006).

The ecolimits then installed in Rocinha had a morphology defined by rails connected to steel cables and were overrun. The ecolimits proposed for Rocinha by Toledo and his team were defined as a *peripheral road system that established the boundaries of the region* (O GLOBO, 09/02/2006) *or as a ring road around the community* that *would serve as a limit to the horizontal expansion of the favela.* Around it would be sports courts and leisure centers. However, Toledo's proposal has been the target of conflicts, which put the proposal to build walls, such as those proposed by Conde in 2004, into play:

"The ring is facing resistance from sectors of the municipal government, which have even proposed building a wall for the same purpose. According to Toledo, the community prefers the ring (...). The idea of a wall is not a good one. Without the participation of the community, committing itself to respecting the eco-limits and containing expansion, it would only serve to support more shacks - Toledo said" (O GLOBO, 31/10/2006).

The *Tijuca Massif Forest* and *Tijuca Forest* categories are marked by the removal of most of their trees and the risk that they would disappear if measures weren't taken. According to Anna Luiza Coelho Netto, coordinator of UFRJ's Geo-Hydroecology Laboratory, *the main causes of the degradation are fires, the removal of vegetation for pasture, landfills and disorderly urban growth,* problems generated by the lack of environmental and housing policies (O GLOBO, 10/08/2006).

Regarding the *Alto da Boa Vista APARU, 13 shantytowns* were growing in a disorderly fashion on its area. Seven of them should be completely removed and put the integrity of the Atlantic Forest at risk, as they were located on marginal river protection strips, on hillsides and because they collected water irregularly. In addition to the environmental damage, these favelas would pose the risk of being joined together (as in the case of Complexo dos Dois Irmao and Vila Alice-Julio Otoni), as evidenced by the intertitle *Comunidades podem se unir em só grande favela* (*Communities may unite into one large favela*) (O GLOBO, 07/09/2006).

From a more general perspective on the ecolimits themselves, in 2006 they continued to be debated in terms of their efficiency and symbolism. According to Anna Luiza Coelho Netto, a researcher at UFRJ, they would not prevent the expansion of favelas. For her, their implementation was a medieval practice *(O GLOBO, 10/08/2006)*. When he was a candidate for mayor of Rio de Janeiro, Eduardo Paes answered a series of questions on various topics formulated by the newspaper O Globo, which were published on September 7, 2006. In the section "Urban Disorder", Paes talks about the ecolimits he had conceived in 2001: for him, physical landmarks (train tracks and steel cables) should prevent invasions from happening. But since they didn't prevent anyone from passing through, they would only work if they were monitored. The candidate is clear about fencing: *I'm against fencing* (O GLOBO, 07/09/2006).

In 2007, the ecolimits continued to be the subject of reports in the newspaper O Globo. In Rocinha, where the landmarks had not been implanted throughout its perimeter (O GLOBO, 03/06/2007), the existing ecolimits were being disrespected (O GLOBO, 05/05/2007). The houses *outside the area demarcated by the eco-limits* (O GLOBO, 03/06/2007) were invading a *preservation area* (O GLOBO, 05/05/2007). In the defense of new urban rules for Rocinha, it is compared to the *formal city*: *like the formal city, new constructions in Rocinha will have to comply with building codes* (O GLOBO, 21/08/2007). New irregular constructions should be inhibited and land use and occupation regulations should be implemented so that the favela can be integrated into the formal city, which would require respect for ecolimits (O GLOBO, 22/08/2007).

Specific localities in Rocinha were also called into question by the limits: some shacks were advancing over the *Cobras e Lagartos* forest, being located outside the ecolimits *(trails interconnected by steel cables), installed by the city hall to delimit the stretches where there can be no construction* (O GLOBO, 23/08/2007). The need to remove these houses or shacks also came up in this discussion (O GLOBO, 23/08/2007, O GLOBO, 28/08/2007 and O GLOBO, 06/09/2007).

More general spatial categories are also being mobilized this year. To avoid damaging the *Atlantic Forest*, eco-limits would have to be implemented in all *communities* (O GLOBO, 03/06/2007). The latter are *growing in Environmental Protection Areas* (O GLOBO, 29/07/2007), while the favelas *are growing wildly, deforesting the hillsides, even in environmental preservation areas* (O GLOBO, 17/12/2007), causing *damage to quality of life, tourism, safety and urban planning* (O GLOBO, 31/12/2007).

Back in 2008, Fernando Gabeira, when he was a pre-candidate for mayor of Rio de Janeiro, committed himself to containing the expansion of the favelas. To this end, the politician proposed that the creation of ecolimits be linked to the construction of pacts with the residents of the communities so that monitoring would take place within the delimited area itself (O GLOBO,

12/03/2008). Candidate Eduardo Paes, on the other hand, defended the need for constant aerophotographic monitoring of the communities associated with the construction of ecolimits (O GLOBO, 23/10/2008). Once elected, Paes guaranteed that ecolimits would be a priority in the environmental sector (O GLOBO, 06/11/2008).

This year's debate was also marked by the defense of the need to contain the disorderly growth of *favelas* on the slopes of the *Tijuca National Park* (O GLOBO, 18/10/2008), in *areas of environmental preservation*, *green areas* and *above the 100th level* (O GLOBO, 04/10/2008). More specific categories were also mobilized, such as *Favela do Pereirão*, which was growing *towards the preserved forest* and *Guinle Park* (O GLOBO, 11/12/2008), and *Dona Marta, a 634-metre wall to establish eco-limits* that would improve the quality of life in the area (O GLOBO, 24/12/2008). In this sense, the delimitation of favelas with eco-limits continued to be seen as a solution to the problems associated with urban disorder (O GLOBO, 17/09/2008), environmental degradation and the city's image (O GLOBO, 05/10/2008).

The players taking part in this second stage of the evolution of the ecolimits device are diverse. These are officials linked to the municipal secretariats for the Environment and Urban Planning and the state secretariat for Urban Development. Vice-governors, state and federal deputies are also active during this period, as well as candidates for mayor of Rio de Janeiro and a Civil Police inspector. In addition to representatives of residents' associations, specialists such as environmental and security researchers, as well as architects, are also invited.

In this long period, which stretches from the proposal to build walls in Rocinha until 2008, the spatial categories mobilized were both general and specific. The ecolimits listed categories such as *green areas*, *environmental preservation areas, areas above the 100-meter level*, *Atlantic Forest, favela reserves, formal city, favelas, poor communities in Rio* and *communities*, which do not refer to precise locations in the city.

More specific categories were also widely mobilized: *Maciço da Tijuca*, *Parque da Tijuca*, *Floresta da Tijuca*, *APA de São José*, *Parque Guinle*, *APARU do Alto da Boa Vista*, *Babilônia*, *Morro Azul*, *Chàcara do Céu*, *Vidigal*, *Favela do Pereirão* and *Dona Marta* are just a few examples. The *Rocinha* category stands out at this point, as it is the subject of particular designs for ecolimits and the subject of discussions about the need for urban planning regulation in favelas, either because of its size or its privileged location in the city of Rio de Janeiro. This becomes even more visible if we take into account the mobilization of categories that describe localities internal to Rocinha, such as *Laboriaux* and *Cobras e Lagartos.*

The ecolimits were debated in different ways during the period described above. Their efficiency was questioned in several situations, in which houses and shacks disrespected the imposed limits or grew vertically. Reformulations of the project, such as those proposed by then mayoral candidates Fernando Gabeira and Eduardo Paes, were based on the belief that the ecolimits policy should be associated with other strategies if it was to be successful.

The appropriate morphology for containing the growth of the favelas was also at the heart of the discussions: according to the periodization described here for the device, this second moment begins with the proposal for a reformulation of the physical dimension of these boundaries. The proposal by the winning architect of the IAB competition, Luiz Carlos Toledo, which consisted of a ring road associated with leisure facilities, constitutes an explicit conflict that has the material dimension of the boundary as its focus. Another content of the debate that surrounded the ecolimits at this time was the meaning of the material dimension itself: the wall was considered a strategy of separation and *apartheid* that denounced the incapacity of the state to solve urban problems.

And there are many urban problems associated with the ecolimits policy. Its implementation was deeply linked to the idea of environmental degradation caused by the growth of slums. This degradation would damage the image of the city itself, since the occupation and destruction of green areas would jeopardize the *beauty of the city*. Although they were justified by containing the growth of favelas on green areas, the ecolimits seem to have been closely linked to the issue of violence. The element that marks the "beginning" of this moment is constituted as a response to illegal activities and the *war for control of the traffic*.

The production of ecolimits was also associated with the disorder, irregularity and informality that takes place in favelas in particular. Disorderly growth also suggested the problem of the fusion or "conurbation" of favelas, either through the occupation of areas that were still preserved or through the strengthening of armed groups within these spaces. In this sense, the issue of integrating favelas or communities into the formal city was also present at this point in the process.

A final aspect refers to the benefits brought to slum dwellers by the construction of physical boundaries. The successful implementation of basic infrastructure in these areas and the reduction of risk situations depended on containing their growth, especially in areas on hillsides or river banks.

6.3. The value of form: the Dona Marta wall, Bill 245/2009 and the Rocinha Ecological Park

On May 8, 2008, the article *Cabral wants to combat the growth of favelas* reported on a meeting held with Governor Sérgio Cabral and his main secretaries to discuss the policy of containing favelas in Rio. On this occasion

proposed *stronger barriers to the growth of communities than the current eco-limits*, which would consist of roads, buildings or walls to contain the growth of *communities.* The idea was to build more robust boundaries than the *current eco-limits*, which are made of wood and steel cables, set up by the city council and located at *50 points around the city*. For Carlos Minc, then state secretary for the environment, these eco-limits failed to contain the growth of the favelas, while the municipal department claimed that most of them were respected by the residents.

The spatial categories mobilized by this news item, an element of the device that marks a new moment in the policy of ecolimits, are *communities, Rocinha, communities next to Tijuca Park, in the North Zone, and Pedra Branca, in the West Zone, favelas where there will be PAC works* and *forest areas*. The *communities* should have their growth contained and *Rocinha* is said to be expanding in the direction of the City Park. According to the report, this favela is a source of concern: *Governor Sérgio Cabral is particularly worried about the expansion of Rocinha* (O GLOBO, 08/05/2008). The *communities next to Tijuca Park, in the North Zone, and Pedra Branca, in the West Zone,* should also have their expansion contained, while in the *favelas where there will be PAC works*, the demarcation will be done by building roads. *Forest areas*, meanwhile, are occupied by the growth of *communities*.

The interface there is between communities and forest areas. However, the environmental motivations are not explicitly stated. The growth of slums is not linked to a more general urban problem: it is, in itself, an issue that must be solved by the state through physical containment.

In the other news items selected for this section, the ecolimits are given different names and a distinction is made between the delimiters already in place under the 2001 project and those that were being proposed or installed in 2008 and 2009. The "old ecolimits" had their physical dimension described by *small concrete pilasters crossed by loose and rusty steel cables* and their 37 kilometers of barriers were intended to protect the forests from new construction (O GLOBO, 24/05/2008).

Some of the columns were already dilapidated and were used as clothesline supports, demonstrating the abandonment suffered by the program (O GLOBO, 24/05/2008). The landmarks had been disrespected and had undergone *drastic modifications* (O GLOBO, 17/01/2009). For former mayor Cesar Maia, the ecolimits installed during his administration were an *experimental project* implemented at random, without any methodological formulation (O GLOBO, 24/05/2008).

Of the 33 favelas where they had been set up, 29 did not have a map locating the demarcations. In 2007, a project was drawn up to assess the situation of the ecolimits, but this was not carried out. Eduardo Paes, previously opposed to the construction of walls (when Luiz Paulo Conde proposed them in 2004), now defends them because he recognizes that the instrument used in 2001 was *fragile* (O GLOBO, 24/05/2008). In this sense, the "old" ecolimits were associated with neglect and lack of supervision, but also with an inability to contain the growth of the favelas.

The "new ecolimits" were called *walls* (O GLOBO, 24/05/2008, O GLOBO, 28/03/2009, O GLOBO, 01/04/2009, O GLOBO, 08/04/2009 and O GLOBO, 22/06/2009b), *barriers* (O GLOBO,

24/05/2008), *belts* (O GLOBO, 17/01/2009), *walls* (O GLOBO, 28/03/2009) and *boundary markers* (O GLOBO, 22/06/2009a). Their objectives were associated with the protection of hillsides and forests (O GLOBO, 24/05/2008 and O GLOBO, 28/03/2009), the limitation of favelas (O GLOBO, 17/01/2009), the separation of favelas from green areas (O GLOBO, 01/04/2009), the containment of the disorderly growth of Rio's favelas (O GLOBO, 08/04/2009), the containment of a concrete threat: the growth of favelas (O GLOBO, 22/06/2009a), the protection of green areas and the fight against "organized crime" (O GLOBO, 22/06/2009b).

The project carried out by the state government through the Rio de Janeiro State Public Works Company proposed the same standard for all the favelas: a concrete wall *based on iron rebar, three meters high.* This pattern had already been thought of as a hollow wall, but was discarded because it was feared that it would be easily destroyed (O GLOBO, 28/03/2009). The design thought up by the city council called for the use of concrete and metal in a 2.60 meter barrier flanked by a road that would surround the favelas, separating the urban part from the part that will be reforested or the virgin forest. According to Carlos Alberto Muniz, then Municipal Secretary for the Environment, it was a *more ostentatious* demarcator that didn't impede the *view of the green areas* (O GLOBO, 17/01/2009). In May 2009, the City Council began to support the construction of the walls, abandoning its proposal (O GLOBO, 24/05/2008).

Journalist Zuenir Ventura criticized the intention to paint the walls green and proposed that the ecolimits *should* be living fences: *to disguise this, the state's Public Works Company intends to paint these walls green or plant ivy next to them. That's a real lack of creativity. If you want to pretend that the concrete is green, why not just put up living fences?* (O GLOBO, 08/04/2009).

These new ecolimits were able to mobilize a great debate that divided opinion. For José Saramago, a Portuguese writer, the three-meter-high reinforced concrete walls could be compared to the walls of Berlin and Palestine, which has also been done on other occasions (O GLOBO, 22/06/2009b). Again according to Zuenir Ventura:

Someone insinuated that this feature is a kind of consecration of the broken city. I hope that wasn't the intention of the creators, but the result could be the same. In a land where everything is transformed into a symbol - the Sugar Loaf, the beaches, Corcovado - these concrete walls will take on an emblematic meaning of the worst kind, evoking examples of bad repute such as the Berlin Wall, or the "Wall of Shame", that of Palestine and that of the USA on the border with Mexico (O GLOBO, 08/04/2009).

For others, this comparison is a *historical impropriety* (O GLOBO, 22/06/2009a). Regina Chiaradia, then president of the Botafogo Residents' Association, believes that the word *wall* suggested Saramago's negative interpretation. In her opinion, the word *ecolimites* should be used instead, and these did not mean a *segregating element*, but *a way of protecting the forest* (O GLOBO, 01/04/2009).

At the moment, the walls are also considered incapable of containing criminal groups, and they are seen as symbols of intolerant action and separation. This measure would have serious repercussions for the image of the city itself:

"Cabral's Wall contradicts the spirit of an inclusive city and associates Rio with a symbol of intolerance at a time when we are bidding to host events (the World Cup, the Olympics) that bring together tolerance and acceptance among different and diverse peoples" (O GLOBO, 22/06/2009b).

One of the articles states that the debate on the walls was marked by political rushes and ill-founded visions of the city, *ideologizations, political interests and opportunism. It also argues that,* very often, the argument was used that the residents of the favelas were against the construction of the wall. However, according to a survey by the DataFolha Institute, *the walls are more pleasing to those who will be inside them than to those who will be outside* (O GLOBO, 22/06/2009a).

Control of the expansion of the favelas was to be done primarily around the Tijuca and Pedra Branca massifs (O GLOBO, 17/01/2009). At the time, twelve favelas were chosen for the construction of the walls. Eleven of them were located in the South Zone of the city and six had not expanded in the period between 2004 and 2008 (O GLOBO, 24/05/2008). The Morro de Santa Marta favela or just Dona Marta was chosen to receive the first of the walls (O GLOBO, 17/01/2009).

The conditions for implementing the ecolimits were to cost R$40 million, from the State Fund for Environmental Conservation (FECAM) (O GLOBO, 24/05/2008). The installation of the walls was also to be accompanied by reforestation efforts in the areas where removals had taken place (O GLOBO, 17/01/2009).

In addition to the announcement of the construction of the new ecolimits and other articles that discussed the issue, the actual construction of one of these walls is an element that integrates this moment of the ecolimits device. The Dona Marta wall concentrated these discussions, which resulted in the project being abandoned after it was finished. As already described, the form of the *wall* linked the project to ideas of segregation, discrimination and separation, which seems to have made it impossible to continue the work in other favelas. This debate seems to have gained even more strength due to the very location of the pilot project. During this period, Dona Marta functioned as a kind of laboratory for urban policies aimed at favelas. In addition, the fact that the favela had been given an inclined plane that acted as a barrier on one of its sides meant that the ecolimit took on the meaning of a total enclosure.

In the elements of the device described above, *Dona Marta* or *Morro Dona Marta* is described as a community that *lost 125 square meters* between 2004 and 2008 (O GLOBO, 24/05/2008) or as a community that has *shrunk slightly in the last ten years, losing 434.35 square meters of area* (O GLOBO, 17/01/2009). It would be the first favela to receive the wall (O GLOBO, 28/03/2009), which would be 900 meters long (O GLOBO, 24/05/2008). In addition to the construction of the wall, Dona Marta was to receive or was already receiving *a series of social, security and infrastructure actions* (O GLOBO, 17/01/2009): *wireless internet signal and special civil and criminal courts, as* well as a *plan to modernize electricity distribution* (O GLOBO, 28/03/2009).

The *Cantagalo* space category is qualified by the fact that the boundaries were relatively respected. *In Pavão-Pavãozinho*, the barriers were ignored (O GLOBO, 01/04/2009). On the other hand, the *21 of the 33 favelas with the barriers, which were flown over by municipal secretaries*, showed horizontal growth and were not targets for inspection (O GLOBO, 24/05/2008). The *13 communities from Urca to São Conrado*, on the other hand, showed an *impressive increase in occupation,* the existence of buildings in areas at risk and beyond the boundary markers. According to the municipal secretaries who carried out the flyover, the new houses and those in risk areas would be removed (O GLOBO, 17/01/2009).

Rio's favelas are described by their disorderly growth. In addition, according to the authorities, their *residents and with the exception of speculators, everyone wants to contain the disorderly advance of Rio's favelas into the surrounding forests* (O GLOBO, 08/04/2009). The favelas, which are growing towards the forest and risk areas (O GLOBO, 22/06/2009a), must be separated from the green (O GLOBO, 01/04/2009). However, there are voices arguing that the construction of walls would not be able to contain any expansion and that its residents could not be held responsible for the violence in Rio (O GLOBO, 22/06/2009b).

The category *"cool"* is mobilized at this time and is characterized by the advance *over the green* (O GLOBO, 17/01/2009). This category can be compared to the *luxury homes* present at an earlier time: although they can't be identified as favelas, they have attributes that allow for a comparison, especially in terms of disregard for urban planning rules and growth over preserved areas.

Other spatial categories are *green*, whose area is *occupied* by *favelas* and *bacana* (O GLOBO, 01/04/2009 and O GLOBO, 17/01/2009), *hillsides* and *forests*, which are invaded (O GLOBO, 17/01/2009), the *forest*, which must be respected (O GLOBO, 01/04/2009) and *environmental preservation areas*, which require preservation (O GLOBO, 22/06/2009a). *Neighboring forests* are related to favelas, but are not qualified by other attributes (O GLOBO, 08/04/2009), as well as *green areas* (O GLOBO, 22/06/2009b).

The interfaces present in these elements can be described as the occupation of "favela-type" categories over "preservation area-type" categories. The *urban part* and the *part that will be reforested or virgin forest are* explicitly related (O GLOBO, 17/01/2009), *favelas* and *green areas*

(O GLOBO, 22/06/2009b), *favelas* and *green*, which must be separated (O GLOBO, 01/04/2009) and *favelas* and *environmental preservation areas*. In this last pair, the relationship between the categories is considered *a harm that affects everyone* (O GLOBO, 22/06/2009a). We can also mention a more specific interface, that between *Dona Marta* and the *forest* (O GLOBO, 28/03/2009).

This third moment that constitutes the temporality of the ecolimits device is also marked by the formulation of a bill in the City Council that is directly related to the reorientation of the Ecolimites project and the construction of the Dona or Santa Marta wall. Bill 245/2009, formulated by then city councillor Leonel Brizola Neto Leonel Brizola Neto, was submitted to the City Council on June 26, 2009. Its aim was to regulate the implementation of ecolimits in the city of Rio de Janeiro.

According to the wording presented by the councillor and attached to the Bill under the subtitle "justification", the initiative is driven by the need to regulate the ecolimits already in place in Dona Marta, about which there has been intense public debate, news has been broadcast, the population has been undecided and because of their permanent repercussions in the city. Despite being approved by all the committees it was submitted to, the bill was not sanctioned by the mayor.

In this document, *the ecolimit aims to prevent the expansion of disorderly urban growth* and *ensure the integrity of the fauna and flora of the Municipality of Rio de Janeiro* (Article 2). Article 6 defines the forms the ecolimit can take: warning signs (location, extent, other information), steel cables, cycle paths, leisure areas. Article 7, which should be subject to a plebiscite, prohibits the construction of ecolimits in the form of walls. The location of these boundaries would be between the reserves and the urban centers (article 6, item II) within the city of Rio de Janeiro. The Bill also defines the conditions for implementation: the ecolimits would have to be approved by the City Council and authorized by the Municipal Department of the Environment upon presentation of the Environmental Impact Study and Environmental Impact Report. Supervision and maintenance would be the responsibility of the Municipal Environment Department.

In this element of the ecolimits device, the spatial category *municipal forest reserves* is qualified because it contains flora and fauna that must be preserved. The *urban center*, on the other hand, shows disorderly growth that must be controlled. The *Dona Marta favela* has its morphology marked by the construction of a three-meter-high wall next to it and the *landscape of Rio de Janeiro* is taking on a different composition because of the construction of such walls. It is interesting to note that the category of *favelas* or similar is only used in the justification and not in the body of the proposed law, even though the ecolimits were designed to contain the growth of favelas. The use of the *urban center* category suggests that the instrument could be applied to any area of the city.

The interface produced in the document relates the spatial categories *municipal forest reserves* and *urban center* to the *ecolimit* discontinuity. The relationship can be described as the disorderly growth of the urban center over reserves which, however, must be preserved because of the fauna and flora they contain. The ecolimits would serve as an instrument capable of reversing this situation of disorderly growth, creating a discontinuity that would inform about the limits of each of the juxtaposed spatial systems.

Ever since the Ecolimites project was announced, Rocinha has been at the forefront: it would be one of the favelas to receive the new walls. Despite protests from the favela's residents, such as a plebiscite that rejected the idea, the state government started work. Shortly afterwards, considering the residents' rejection and the intense public debate that had been produced by the construction of the Dona Marta wall, the government backed down and subjected the project to significant alterations, which resulted in the Rocinha Ecological Park.

On April 29, 2009, the article entitled *Almost 50 houses 'pierced' Rocinha's ecolimit* announces the construction of a new park for Rocinha. This element deals with the boundaries of the *wall* and *the ecolimit*. The three-kilometer wall, the new ecolimit, was intended to *preserve the area of Atlantic Forest that still exists there, ratifying the ecolimits that had already been made.* In order to carry out the works, approximately 400 families had to be relocated. icaro Moreno, president of EMOP (the public company responsible for the project), said that the works would bring benefits to the favela, such as a new leisure area, improved circulation and the regularization of housing.

However, the news report highlights that another proposal was advocated by residents. The text describes the alternative, according to the president of the Rocinha Residents' Association, Antonio Ferreira: *the idea would be to build a ring road starting from the wall of the Zuzu Angel Tunnel, integrating Rocinha, Alto Gâvea and Estrada das Canoas, going around the favela. This ring road would be completed with an ecotrail: a paved hiking trail, with benches and toys for children at some points* (O GLOBO, 29/04/2009). This is a clear dialog with the project conceived by Toledo when he drew up the Rocinha Master Plan as part of the IAB ideas competition. His design is brought from one moment to the next in order to confront the state government's proposal. As in the "second moment", the ring road project dialogues and conflicts with the wall form for the ecolimits.

A few days earlier, a plebiscite had been held on the construction of the walls and, out of 1,056 votes, only 50 were in favor of it. The result, however, would not be taken into account by the state government. According to Ferreira, the ecolimits had worked in the favela: "*The ecolimits worked well here, what exists there is precarious housing. The population is against the wall, which is a symbol of division*". However, according to the report, "*some of the rails used to demarcate the area have been removed and replaced elsewhere. And a steel cable from the ecolimit is even used as an improvised swing by the children*" (O GLOBO, 29/04/2009).

The following month, a new report revealed Governor Sérgio Cabral's decision: *after stating that he would not change the original design of the Rocinha wall, Governor Sérgio Cabral went back on his word and accepted a suggestion from residents to erect walls in some parts of the ecolimit.* The new project would include 60-centimeter walls and a railing in areas of the ecological park, as well as a bike path, also suggested by residents. The text makes direct reference to Luiz Carlos Toledo, stating that the residents' suggestions were included in the project developed by the architect at the Association's request. According to its president, *(...) the wall and the ecotrails represent the integration of people and free access to the forest at specific points. The project maintains our freedom to come and go. The wall, on the other hand, gives the feeling of a ghetto* (O GLOBO, 21/05/2009).

In the article entitled *Parque na Rocinha terá 20 itens para lazer de moradores: projeto prevê muretas no lugar de parede em algumas trechos,* Rocinha's ecolimits are described as being intended to contain the community's expansion towards the Atlantic Forest. The text reported that the design of the new park had already become a consensus between favela leaders and EMOP:

"During the presentation of the project to the press, both the president of the State Public Works Company (Emop), Icaro Moreno Júnior, and the president of the Rocinha Residents' Association, Antonio Ferreira de Mello, or Xaolin, and the president of the Popular Movement of Favelas, William de Oliveira, said that the discussion about the wall is over. According to them, the project has the approval of the residents" (O GLOBO, 26/06/2009).

According to Anderson Café, the EMOP architect responsible for the project, in an exploratory interview conducted in June 2012[44] , his team designed a project that was denied by the governor due to the lack of participation by the residents. As a result, the residents of Rocinha were consulted through the Oficina do Imaginàrio, a meeting that mobilized people "from 4 to 80 years old", according to the architect. In this workshop, based on a model brought from Bogotà by the team itself, residents were asked to design the facilities they wanted to see built in the new park. According to Café, ninety percent of the requests were met, even those that were not part of the ecological park concept, such as courts, barbecue pits and an amphitheater, and by that time, half of the park was ready.

According to the architect responsible for the project, the park could be described as a linear form of approximately eight square kilometers in area and 900 meters in length. The park was located where the *Cobras e Lagartos* shanty town had previously stood, between Rua Dioneia and Laboriaux, a precarious place, *the* focus of diseases such as tuberculosis (*the highest incidence of tuberculosis in the world*), with little lighting, taking advantage of the road produced by its residents. The architect added that the area is within the *buffer zone of the Tijuca National Park.*

In order for the park to be built, around a hundred houses had to be removed. For Café, there would

44 The interview was conducted at EMOP's headquarters by the author.

still have to be a removal, which was the subject of a legal dispute. The work, in which 22 million reais were invested from FECAM or oil royalties, was halted for a year due to a lack of funds, which were redirected to emergency work. At that time, no body had yet been set up to manage the Rocinha Ecological Park. According to EMOP's architect, the Secretariat of Culture and SUDERJ had already been *approached*. The management of the park was like a *game of push and shove*, especially until the arrival of *the pacification forces*. Negotiations were also being made with the managers of PARNA Tijuca, due to the fact that the Ecological Park is in its buffer zone.

The spatial categories mobilized through the interview with EMOP's architect and the news in O Globo about the construction of *the* Rocinha Ecological Park, an event that organizes and makes up the third moment of the device described here, are: *Rocinha*, *Rocinha Ecological Park*, *Cobras e Lagartos, Tijuca National Park buffer zone, forest area*, *Atlantic Forest* and *green areas*.

Rocinha is qualified by the fact that it expands over *green areas*, *forest* and *Atlantic Forest*, as well as being one of the places that would receive the new walls. *Cobras e Lagartos* was located in a clearing next to the forest, in the buffer zone of the *Tijuca National Park*, and had wooden dwellings. However, it was already *possible to see a number of trucks carrying construction material passing through* (O GLOBO, 29/04/2009). As well as signifying an area of favela expansion, this category is marked by ideas such as precariousness and unhealthiness.

The *Rocinha Ecological Park* can be described as a linear park located between Rua Dioneia and Laboriaux, at the top of Rocinha, with approximately 8,000 km^2 . The project included gym equipment, an amphitheater, a children's playground, courts, barbecue pits, a forest and a bike path. It meant a leisure space for Rocinha and was supposed to act as a boundary, preventing the favela from expanding into forest areas.

The interfaces mobilized by this ecolimit associated with a park relate *Rocinha* or its specific localities, *Cobras e Lagartos*, *Portão Vermelho* and *Laboriaux*, to the categories of *forest area*, *Atlantic Forest* and *green areas*. The former would expand over the areas occupied by the latter, which should nevertheless be preserved. It is worth noting that, for the residents, the installation of low walls would allow Rocinha residents free access to the forest, which would result in a different relationship between the mobilized categories.

Considering the three subdivisions[45] of the third moment of the device-ecolimites, a wide variety of actors took part. They included the governor, the state secretary for the environment, the former mayor and mayoress, the municipal secretary for the environment and city councillors, as well as

45 The three subdivisions are: a) announcement of the walls and construction of the Dona Marta wall; b) Bill 245/2009; c) construction of the Rocinha Ecological Park.

EMOP officials. Also taking part were journalists, writers, intellectuals, residents of the favelas being intervened in and their residents' associations.

When the construction of the new walls was announced and the project was publicized, the spatial categories mobilized were both general and specific. They dealt with *communities, favelas and forest areas* in general, as well as the *eleven favelas* where the new boundaries would be built, the *thirty-three favelas* where the old ecolimits had been built, *Cantagalo* and *Pavão-Pavãozinho*. *Dona Marta* stood out during this period, as it was the first slum to receive the wall and was where most of the favela policies were implemented.

Bill 245/2009, on the other hand, used very general categories, such as *urban centers* and *reserves*. The generality of *urban centers* means that the text does not deal directly with favelas, although Dona Marta is mentioned in the bill's justification. In this way, it suggests that the instrument could be applied to any part of the city that is in contact with municipal reserves, preferably.

Taking into account the debate surrounding the construction of Rocinha's ecolimit, the spatial categories are considerably more specific. The categories mobilized relate specifically to *Rocinha* and its localities, such as *Cobras e Lagartos, Portão Vermelho, Portão Vermelho and Laboriaux.*

This third moment was deeply marked by a comparison between the old and new ecolimits, which meant a reconfiguration of the policy with a view to making the instrument more efficient. It should also be noted that this moment was capable of producing an intense public debate in which different voices participated, despite the "little materiality" that was actually built. After all, only the walls of Dona Marta and Rocinha were erected.

In general, the elements that make up this moment of the device are associated with different urban problems. The first is the environmental degradation caused by the expansion of slums over green areas. In this sense, the ecolimits are intended to contain the growth of slums in order to guarantee the preservation of areas of environmental value in the city. Another issue associated with ecolimits is urban disorder. The disorderly growth of favelas would be a symptom of irregularities and disrespect for urban planning rules. The state should therefore ensure that limits and laws are complied with, just like in any other area of the city. In this way, the other interventions carried out in the favelas and the very construction of the ecolimits would act to integrate them into the *formal city*. However, the moment was significantly marked by the discussion of the meanings of the physical dimension of the ecolimits, bringing walls full of symbolism into the debate on socio-spatial segregation and the discontinuities between the asphalt and the favela in the city of Rio de Janeiro.

6.4. Ecolimits in the 2011 Master Plan

Considering the Master Plan as an important instrument in the definition of public policies and its formulation as a prominent arena for the debate of interests relating to urban policy in the city of Rio de Janeiro, some documents that mark its evolution are constituted in this work as elements of the ecolimits device. This is because the physical delimitation of favelas/green areas was a relevant theme in the conception of these documents, explicitly included in articles and sections.

In this sense, we consider the 2011 Master Plan for Sustainable Urban Development of the Municipality of Rio de Janeiro as a moment in the debate about ecolimits. This moment is a consequence of the evolution of the city's Master Plan and the previous moments described above. The production of this Master Plan is very complex, combining different agents and power relations, and is also marked by a time span of approximately 20 years. As our intention here is not to describe the particular process that gave rise to this document, nor to describe the evolution of the ecolimits simply by relating them to the Master Plan, we have chosen to identify documents of great importance in its processing. The following documents will be considered: the Complementary Bill n° 25/2001 (2001), the Complementary Bill n° 25/2001 - Substitute n° 3 (2006), the Annex to Message n° 36/2009 of the Complementary Bill n° 81/2001 (2009) and the Sustainable Urban Development Master Plan for the Municipality of Rio de Janeiro - Complementary Law n 111/2011 (2011). Below is a brief description of these elements.

In 1992, Complementary Law no. 16, the Ten-Year Master Plan for the City of Rio de Janeiro, was instituted as a basic instrument of the municipality's urban policy. In 2001, the City Statute (Law 10.257 of 2001) was approved, which, among other stipulations, required that master plans be reviewed every ten years. In compliance with the federal law, the Municipal Executive sent a proposal for revision to the City Council, the Complementary Bill° 25/2001.

"The difficulty in carrying out this task with a technical team made up exclusively of lawyers, and the discontent caused in the Legislature and among technicians in the Executive itself, corroborated by the Public Prosecutor's Office in response to popular pressure spearheaded by the Forum for Monitoring the Master Plan, meant that the Legislature suspended consideration of the first revision text and sent it back to the Executive, where it was addressed to technicians in the Urban Planning Department to consolidate the revision. The original document was then fragmented into various themes and handed over to the different sectors responsible for the municipality's urban management system" (IPPUR, 2010: 10-11) .[46]

46 In 2009, the Rio de Janeiro City Council, through the Special Commission on the Master Plan, asked the Institute for Urban and Regional Planning and Research of the Federal University of Rio de Janeiro (IPPUR/UFRJ) to analyze the proposals for revising the Master Plan that were underway at the time. In January 2010, the technical evaluation report of the proposals to revise the Ten-Year Master Plan for the City of Rio de Janeiro was published (IPPUR, 2010). The document consists of an assessment of the proposals to revise Rio de Janeiro's DP based on the 1992 Master Plan, Substitute No. 3 and the Annex to Message No. 36 (2009). It is structured around a specific assessment of the themes of

In 2005, a working group led by the Municipal Urban Planning Department was set up within the Municipal Executive to coordinate and monitor the process of revising and amending the Master Plan. Based on the work of this group, the revision proposal called Substitutivo n° 3 or Mensagem n° 81/2001 (based on Bill n° 25/2001) was sent to the City Council in 2006. This document represents the third consolidated proposal by the public authorities to modify the document that proposed the revision of the City's Master Plan, drawn up in 2001. It stood out from the other substitutes and formed an important basis for the production of the text approved in 2011.

In the same year, the Special Commission for the Revision of the Master Plan was created within the City Council, which organized the Seminar "Master Plan: the challenge of the new legal-urban order" in order to provide parliamentarians with subsidies for the task of revising the Plan in question.

"Dissatisfaction with the text of Substitute 3 has gained the support of technicians and scholars, even reaching sectors of the executive itself, as demonstrated by the various attempts to paralyze the process by representatives of civil society, and the forwarding of the revision proposal in what interests it, consolidated in the 44 amendments in the Annex to Message No. 36" (IPPUR, 2010: 11).

In 2009, a new commission was set up by the municipal executive in order to adapt the proposal in Substitute No.° 3 to the changes in legislation (PREFEITURA DA CIDADE DO RIO DE JANEIRO, 2009). Staff linked to the City Hall took part in this process, and later proposals formulated by councillors involved in the Master Plan's progress in the Chamber were added, which gave rise to the Annex to the Message mentioned above (this is Message 81/2001, Substitute 3). Finally, in February 2011, the revision of the Master Plan was approved (Complementary Law° 111/2011), which was renamed the Sustainable Urban Development Master Plan for the Municipality of Rio de Janeiro.

In the 2001 text (Complementary Bill No.° 25/2001), the following *limits* can be identified: *limits for urban expansion*[47] and *physical demarcation of the limits of the Favela expansion area*[48] . For the first boundary, the aim is to control occupation of the hillsides. For the second, the objectives are: to stop the process of over-densification of the Favelas and the exploitation of real estate in the rental of properties in the communities, and to prevent the construction of new buildings of any kind beyond the established limits. For both there is no definition of the physical dimension, but for the second it is clear that the intention is to set a limit which in itself has its own morphology. The *limits for urban expansion* are located on hillsides and their setting must take into account *the specific nature of each area and the risks of landslides or mudslides identified.* According to the title of the Subsection that

---

housing, the environment, transport, democratic management and macro-zoning, as well as a more general assessment of the structure and content of the proposals.

47 Title VII: Sectoral policies; Chapter I: Environmental policy and enhancement of cultural heritage; Section IV: Programs; Subsection III: Program to protect slopes and lowlands subject to flooding, Art. 137, alinea I

48 Title VIII: Final provisions; Art. 238, par. 3, subparagraph II

describes it, this is an action within the scope of the *Program for the Protection of Slopes and Lowlands Subject to Flooding.* The *physical demarcation of the boundaries of the Favela expansion area* is more specifically located in favelas and their expansion areas. In this document there is no mobilization of the signifier "ecolimits".

In Substitute No.° 3, of 2006, the term *eco-limits* is used to *mark the boundary between favelas and protected green areas*[49] and, according to Annex III, they should *promote environmental protection.* In this same Annex, *Guidelines by Occupation Macrozone*, eco-limits are only mentioned in the part dedicated to the Controlled Occupation Macrozone, which indicates that eco-limits should only be implemented in this Macrozone. It also uses the term *physical limits*[50] , which are designed to *contain* the *growth and expansion of slums* and must be in line with the establishment of special urban planning rules in order to contain slums. The text also deals with *the limits established for urban regeneration areas*[51] , which must be respected in the city's occupation and growth processes.

In the Annex to Message No.° 36, published in 2009, the *Eco-limits* constitute the *boundary between occupied areas and those that are protected or have vegetation cover of any kind*[52] and should curb *irregular urban expansion over areas that have vegetation cover of any kind.*[53] As in the text of Substitute No.° 3, the eco-limits should be set up in the Controlled Occupation Macrozone in order to *promote environmental protection*[54] . Their installation must be associated with *control mechanisms that guarantee compliance with the physical boundaries* .[55]

In this document, as well as in Substitute No.° 3, the *limits established for the areas restricting urban occupation* are described with the same characteristics and in the same article. The *100-meter quota* is also introduced, *which* is intended for the *conservation and environmental recovery of the Atlantic Forest Biome*[56] . This limit defines the areas above the quota in which the opening of roads is prohibited, as well as the implantation of allotments or armaments of private initiative[57] . Another limit described is the level *of sixty meters*[58] which determines that interventions in areas above this

49 Title II: Land use planning; Chapter I: Land use and occupation; Section II: Urban occupation; Art. 11, item VII
50 Title I: Urban policy; Chapter I: The principles and guidelines of the municipality's urban policy Art. 3, item VI
51 (Title II: Land Use Planning, Chapter I: Land Use and Occupation, Section II: Urban Occupation, Art. 9)
52 Title II: Territorial Planning, Chapter I: Land Use and Occupation, Section II: Urban Occupation, Art. 11, para. 1, item VII
53 Title V: Strategies for Implementing, Monitoring and Controlling the Master Plan, Chapter VI: The Land Use and Occupation Control System, art. unnumbered, item V
54 Annex III: Guidelines by Macrozone of Occupation
55 Title V: Strategies for Implementing, Monitoring and Controlling the Master Plan, Chapter VI: The Land Use and Occupation Control System, art. unnumbered, item V
56 Title III: Urban Policy Instruments, Chapter I: General Urban Regulation Instruments, Section II: Land Use and Occupation Law, art.34, par. 1, item I
57 Title III: Urban Policy Instruments, Chapter IV: Environmental and Cultural Management Instruments, Section I: Environmental Management Instruments, Subsection V: Environmental Control and Monitoring, art. unnumbered, par. 1 and 2
58 Title III: Urban Policy Instruments, Chapter IV: Environmental and Cultural Management Instruments, Section I: Environmental Management Instruments, Subsection V: Environmental Control and Monitoring, art. unnumbered

level *throughout the municipality must take into account environmental, landscape and geotechnical restrictions.* Also in the Annex to Message No.° 36 is the *georeferenced physical* boundary[59] . Its implementation is one of the *structuring actions relating to the protection of the Atlantic Forest Biome* and aims to *protect the Atlantic Forest and other areas of environmental importance.*

Complementary Law No.° 111/2011, which establishes the 2011 Master Plan, once again presents the boundary known as the *ecolimit*[60] . There is no difference in content and form regarding ecolimits if we compare the texts of this law and the text of the Annex to Message n° 36, which includes their location in the Controlled Occupation Macrozone. The only difference is in the article numbering, which, however, is in the same Titles, Chapters and Sections. The same is true for the *georeferenced physical boundaries*[61] (there is only a difference in the subsection and article numbering) and for the *physical boundaries* (the difference is in the subsection numbering).

Like Annex° 36, the 2011 text also describes the *100-meter quota* limit, which is aimed at the conservation and environmental recovery of the Atlantic Forest Biome. However, the most recent text does not deal with restrictions on the opening of roads, allotments and streets. The *limits established for the areas of restriction to urban occupation* and the *sixty-meter quota*, present in 2009, are not included in the most recent text.

Regarding the spatial categories mobilized by the boundaries constituted in the elements that make up this moment of the device, we have that, in the 2001 text, the categories directly mobilized by the *physical demarcation of the boundaries of the favela's expansion area* are: the *favela's expansion area* and *the forested area surrounding the favela.* The *expansion area* of *the favela* must have its boundaries physically demarcated and, beyond the limits of this expansion area, no building of any kind will be allowed. The[62] *favelas* are characterized by precarious urban infrastructure and public services and values associated with irregularity and illegality. Their morphology is marked by *narrow streets with irregular alignments, irregularly sized and shaped plots and unlicensed buildings that do not comply with legal standards.* These are predominantly housing areas occupied by low-income populations, which should be urbanized and regularized. On the other hand, the *forested area around the favela* is susceptible to invasion and should be the subject of an *invasion containment and prevention system.*

Considering the *expansion area of the favela* and *the forested area surrounding the favela, we see*

---

59 Title IV: Sectoral Policies, Chapter II: Environmental Policy, Section III: Structuring Actions, Subsection V: Protecting the Atlantic Forest Biome, art. unnumbered, item VII

60 Title II: Land use planning; Chapter I: Land use and occupation; Section II: Urban occupation; art. 15 par. 1, item VII

61 Title IV: Sectoral public policies; Chapter II: Environmental policy; Section III: Structuring actions; Subsection V: Protecting the Atlantic Forest Biome; Art. 178, item VIII

62 Title VII: Sectoral policies; Chapter II: Housing policies; Section IV: Programs; Subsection I: Slum upgrading and land regularization programs; Art. 153 and 154

that these spatial systems are situated on the external limits of the favelas and are juxtaposed. The interface produced by the boundary described above puts the favelas in relation to both their area of expansion and the surrounding forested area: the growth of the favelas creates its own area of expansion and puts pressure on the surrounding forested area. It is clear that other categories associated with this boundary and the categories mobilized are described in the document. These are: *areas destined for environmental protection, Tijuca Massif, Tijuca National Park, area of outstanding landscape value, environmental conservation unit, urban occupation* and *hillside occupation*.

In Substitute No.° 3, *eco-limits* mobilize the categories of *slums* and *protected green areas*. The morphology of slums is marked by growth and expansion, which must be contained *through the establishment of physical boundaries and special urban planning rules*[63] . They must be *restored to environmental conditions, infrastructures put in place and the health and living conditions of the dwellings improved*[64] . They do not form part of the formal fabric of the city, but must be modified in their urban form in order to become part of it. These are predominantly housing areas, *characterized by clandestine and low-income occupation*[65] . Compared to the 2001 text, favelas have gained the attribute of clandestine occupation.

Protected green areas, on the other hand, are *areas outside the eco-limits where* no construction is allowed. They must be protected and are defined as *areas deemed unsuitable by the municipal administration*[66] . The interface between *slums* and *protected green areas*, categories related by the eco-limits, is characterized by the possible expansion of slums over the protected areas. Clandestinely and informally, favelas grow with new constructions and put pressure on the areas outside the eco-limits, where buildings are not allowed.

According to the technical report produced by the Institute of Urban and Regional Planning, Substitute 3 is marked by a "naturalism" that conceives of the environment as a value in itself. This would result in a negative valuation of slums and low-income housing estates:

"This conceptual option, to a certain extent, is what would legitimize the definition of a new 'environmental evil' no longer located in industrial activities, traffic, inadequate disposal of solid waste, atmospheric pollution and other related problems, but in urban occupations considered 'predatory', especially slums and low-income housing developments" (IPPUR, 2010: 59).

Both in the Annex to Message no.° 36, of 2009, and in the Sustainable Urban Development Master

63 Title I: Urban policy; Chapter I: The principles and guidelines of the municipality's urban policy, Art. 3, item VI
64 Title IV: Sectoral public policies; Chapter IV: Housing policy; Section I: Objectives, Art. 150, point V
65 Title III: The instruments of urban policy; Chapter VII: The policy of urban and financial regularization; Section III: Procedures, Art. 174, § 3°
66 Title II: Land Use Planning; Chapter I: Land Use and Occupation; Section II: Urban Occupation; Art. 11

Plan for the Municipality of Rio de Janeiro, of 2011, the *ecolimits* explicitly mobilize the spatial categories of *occupied areas*, *irregular urban expansion* and *areas destined for environmental protection or which have vegetation cover of any kind.* In these texts, the *areas occupied or committed to occupation* are described as those in which use and occupation must be regulated by *limiting densities, the intensity of construction* and *economic activities, depending on the capacity of the infrastructure, the transport network and accessibility and the protection of the natural environment and urban memory*[67] . *The* 2011 text adds *the right to enjoy the natural landscape of the city and the quality of the urban environment* .[68]

These *occupied areas* are not characterized in terms of regularity, informality or legality. In this sense, the ecolimits would be aimed at any type of occupied area, *urban occupation* or *area actually occupied,* and not specifically at slums, as in the text of Substitute No.° 3. It has to be considered that this change in the spatial categories mobilized greatly modifies the content of the limit itself called an *ecolimit.*

Areas intended for environmental protection or which have vegetation cover of any kind must be protected with the help of physical delimiters and mechanisms to ensure compliance. They can be associated with other spatial categories such as those described below. *Areas subject to environmental protection* are those with environmental characteristics that must be preserved from urban occupation. The *Environmental Conservation Zone* has natural, cultural and/or landscape characteristics that are relevant to preservation. These are areas located above elevation 100, buffer zones and *fragile lowland and hillside areas and their associated biomes, which are not occupied or urbanized.* It can be totally or partially transformed into a Nature Conservation Unit and has attributes that justify its preservation. *Nature Conservation Units* are delimited at the time of their creation, which is seen as a basic instrument for protecting the municipality's environment. *Permanent Preservation Areas must be recovered as a priority through the implementation of recovery and revegetation programs.* Their establishment is also considered a basic instrument for protecting the municipality's environment. *Areas unsuitable for urban occupation* constitute the *natural environment* and must be preserved from urban occupation. Just as forests *and other areas with vegetation cover* are an element that conditions urban occupation.

The boundary entitled *physical limits,* present in the Annex to Message No.° 36 and in the 2011 Master Plan, has the function of containing the growth of slums. In this sense, it differs from the ecolimits in that it mobilizes the spatial category of *slums.* These are characterized by precarious infrastructure and public services, narrow streets, irregular alignment and unlicensed construction.

67 Title II: Territorial Planning, Chapter I: Land Use and Occupation, Section II: Urban Occupation, Article 10
68 Title II: Land use planning; Chapter I: Land use and occupation; Section II: Urban occupation art. 14

Their growth and expansion must be contained *by setting physical boundaries and establishing special urban planning rules*. They are predominantly housing areas, characterized by clandestine and low-income occupation (as well as Substitute No.° 3, of 2006). They should be integrated into the formal areas of the city, unless they are in a situation of risk or environmental protection. In the 2011 text, the information that all of the city's tourist potential should be used is added to the category.

In general, it is possible to say that the agents participating in this moment of the device-ecolimits are those who took part in the debate on the revision of Rio de Janeiro's 1992 Ten-Year Master Plan. They are mayors, secretaries of Urban Planning, Environment, Works, Housing, etc., technical staff from the City Hall. They are councillors, members of special committees linked to the formulation of the Plan or not. They are citizens who took part in the hearings, lectures and demonstrations that brought the revision to a standstill, as well as representatives of the most diverse civil society associations.

To summarize, with regard to *ecolimits* in the evolution of the Master Plan, which culminated in the approval of the 2011 text, it can be said that the term is used from the 2006 document onwards, when the project had already been implemented in various situations, and is used until 2011. In the 2001 document, although they had already been announced a few months earlier, ecolimits do not appear as a category. There are other limits used (*limits for urban expansion*, which deals with occupations on hillsides and the *physical demarcation of the limits of the Favela expansion area*) which, in some way, already indicate the interfaces on which the ecolimits will also operate. It should be noted that the issue of favela expansion appears more clearly in the *final provisions of* the Bill.

In the 2006 text, ecolimits make up the interface between *favelas* and *green areas*. In the 2009 and 2011 documents, the *slum* category is no longer mobilized directly by the ecolimits, but is replaced by *occupied areas*. In turn, *green areas* are replaced by *areas destined for environmental protection or with vegetation cover of any kind.* The boundary that mobilizes the *favela* category in these last two elements (2009 and 2011) is called *physical boundaries*, the same one already used in the 2006 Substitute, with the same attributes.

The spatial categories mobilized by the interfaces produced at this point in the device are endowed with significant generality. They do not deal with specific areas in the city of Rio de Janeiro, although they are identified and qualified in terms of their morphologies, uses and associated actions and values. The most specific category mobilized directly by the ecolimit is the Controlled Occupation Macrozone, which corresponds to a large portion of the municipal territory, made up of Planning Areas 1 and 2. In an attempt to classify the mobilized categories described above into large groups, it is possible to say that they are divided into those that should be contained and those that, because they have relevant environmental attributes, should be preserved.

At the moment, the ecolimits are talking about important issues for the city of Rio de Janeiro. The growth of favelas is associated with the threat to the urban environment, the degradation of areas that offer environmental services to the city, the risk to the favela itself when it occupies unstable land or land that is severely deforested, and the loss of quality of life due to the reduction of open spaces and green areas. In addition to these "environmentalist" relationships, the growth of favelas is strongly linked to problems of "urban disorder". The favela, irregular, illegal, informal, precarious, poor, clandestine, disordered, as it grows, extends to other areas of the city the forms, uses and values that organize it as a spatial category.

Final considerations

By taking boundaries not just as segregation markers, but as elements that operate in the relationships between spatial systems, it is possible to understand them in a broader perspective that encompasses other spatial interactions. The act of limiting is an act of differentiation, whether based on morphology, uses or the meanings attributed to the spatial cut-outs it creates. Boundaries act to re-elaborate these dimensions and, therefore, the very categories they relate to.

By considering ecolimits as discontinuities, they become able to participate in relationships that are not reduced to the dynamics of socio-spatial segregation in Rio de Janeiro. They come to act as regulatory mechanisms that organize our experience of the world and the city. They are products engendered in specific situations and moments and related to different visions of the city.

Considered as a device, these boundaries make and unmake interfaces by mobilizing and requalifying spatial categories. The documents described here relate to each other, affect each other and meet in order to respond to the urgent need to manage the city's growth. The device-ecolimites described here as a network between the selected elements is not a total device, comprising all the discourses and practices that interfere in the production of instruments to contain urban expansion in the city of Rio de Janeiro.

However, the construction of other instruments or the consideration of other elements in these instruments is a possibility for further research. The aforementioned instruments of urban planning regulation, such as the city's Building Code, or environmental management, which include the creation of nature conservation units, bring new content to the subject in question and can be thought of as other possible relationships with the elements described here.

Understood as a process, a non-linear evolution made up of moments and enunciations/elements, the device-ecolimites produces generalizations and spatial specifications in the composition of interfaces and, therefore, different interfaces. Sometimes it relates "green areas" and "human occupation", sometimes it deals with the Laboriaux Iocality in Rocinha and the Tijuca National Park Buffer Zone. Sometimes it takes the form of a three-meter high concrete wall with a total length of eleven kilometers in the South Zone of the city, sometimes it is rejected by public opinion and is associated with an ecological park with equipment chosen by Rocinha residents.

If, on the one hand, it deals with general issues such as environmental preservation, on the other, it is localized and needs to be built in specificity, through a negotiation situated with the places, their environmental conditions, disputes and agents. In this way, both the conception of a general idea for the project and its models for dissemination, as well as the specific design for a favela and the treatment of the issue on the ground, are elements that make up the network of the device.

Throughout the moments presented, ecolimits are associated with various urban problems and visions of the city. They mobilize environmental issues, planning and regularization of the city fabric, as well as differentiations between spatial categories in terms of their morphologies, their appropriate uses and actions and their meanings. The production of ecolimits - whether as an idea/intention/project or as a materialized form on the ground - incorporates or is marked by an intention to incorporate the favela, or informal city, into the formal city. This intention is based on the recognition of the significant differences between categories such as favela, community, city and formal city.

Despite the little materiality produced, especially in the 2009 project, the ecolimits created a vigorous public debate. Designing, announcing or building boundaries in the territory, whether in specific favelas or in the fabric of the city in general, has summoned and still summons different opinions and values about what the city is and what it should be.

Bibliographical references

ACSELRAD, Henri. Discourses of urban sustainability. Brazilian Journal of Urban and Regional Studies, n. 1, 1999, p. 79-90.

AGAMBEN, Giorgio. Qu'est-ce qu'um dispositif? Paris: Payot et Rivages, 2006.

AGAMBEN, Giorgio. What is a device? Sociológica, ano 26, n. 73, 2011, p. 249264.

BARNIER, Véronique and TUCOULET, Carole. Avant-propos. In : Ville et environnement : de l'écologie urbaine à la ville durable. Problèmes politiques et sociaux. n. 829, 1999.

BENGSTON,D. FLETCHER, J. and NELSON,K. Public policies for managing urban growth and protecting open space: policy instruments and lessons learned in the United States. Landscape and Urban Planning n. 69, 2004, p. 271-286.

BENGSTON, D. AND YOUN, Y-C. Urban Containment Policies and the Protection of Natural Areas: The Case of Seoul's Greenbelt. Ecology and Society, v. 11(1), n. 3.

Available at: http://www.ecologyandsociety.org/vol11/iss1/art3/

BESSERMAN, Sérgio and CAVALLIERI, Fernando. Technical note on the growth of the favela population between 1991 and 2000 in the city of Rio de Janeiro. Coleçao Estudos Cariocas. Rio de Janeiro, 2004.

BRAZIL. Law No. 4.771 of September 15, 1965. Establishes the new Forest Code.

Available at http://www.planalto.gov.br/ccivil_03/leis/L4771.htm. Accessed on 07/08/12.

BRUECKNER, Jan. Urban Sprawl: Diagnosis and Remedies. International Regional Science Review, v.23, n.2, 2000, p.160-171.

BRUNET, Roger. The phenomena of discontinuity in geography. Mémoires et documents, n. 7, 1967.

BRUNET, R., GRASLAND, C. and FRANÇOIS, J-C. Discontinuity in geography: origins and research problems. L'Espace Géographique, n.4, 1997, p. 297-308.

CALDEIRA, Teresa. City of walls: crime, segregation and citizenship in Sao Paulo. Sao Paulo: Edusp, 2000.

CAMARANO, Ana Amélia et al. Demographic trends in the municipality of Rio de Janeiro. Carioca Studies Collection. Rio de Janeiro, 2004.

CAMARGO, Jean Carlos. Ecolimits or Socio-limits: from "environmental preservation" to socio-spatial segregation?, 2012. Available at: http://observatoriodasmetropoles.net /index. php? option =com_k2&view=item&id=147:ecolimites-ou-s%C3%B3cio-limites ? &Itemid=165 &Lang =en.

Accessed on 03/02/12.

CANÇADO, Wellington. The wall: "ecolimits" and the favelas of Rio de Janeiro. *Minha Cidade,* Sao Paulo, n.09.106, Vitruvius, 2009. Available at : <http://www.vitruvius.com.br/revistas/read/minhacidade/09.106/1854>.

CAPEL, Horacio. The Morphology of Cities: I. Society, culture and urban landscape. Barcelona: Ediciones de Serbal, 2002.

CARDOSO, Adauto L. and ARAUJO, Rosane L. de. The slum urbanization policy in the municipality of Rio de Janeiro. In: CARDOSO, Adauto L (coord.) Social Housing in the Brazilian Metropolis: an evaluation of housing policies in Belém, Belo Horizonte, Porto Alegre, Recife, Rio de Janeiro and Sao Paulo at the end of the 20th century. Porto Alegre: ANTAC, 2007.

CASTRO, Edgardo. Michel Foucault's vocabulary: an alphabetical tour of his themes, concepts and authors.Universidad Nacional de Quilmes, 2004.

CAVALLIERI, Fernando and LOPES, Gustavo. Favelas cariocas: comparison of occupied areas - 1999/2004. Carioca Studies Collection. Rio de Janeiro, 2006.

COMPANS, Rose, A cidade contra a favela: a nova ameaça ambiental. Revista Brasileira de Estudos Urbanos e Regionais, v. 9, n. 1, 2007, p. 83-99.

COSTA, Heloisa. Sustainable urban development: a contradiction in terms? Revista Brasileira de Estudos Urbanos e Regionais, n.2, 1999, p. 56-71.

DE CARLI, Natàlia and HUMANES, Mariano. An in-human ecolimit: fear and social- spatial segregation. 4th Global Conference Fear, Horror and Terror at the interface, 2010. Available at http://www.inter-disciplinary.net/at-the-interface/evil/fear-horror- terror/project-archives/4th/session-6-fht-in-spaces-exterior-and-interior/ Accessed on 05/04/2012.

FERREIRA, Alvaro. Favelas in Rio de Janeiro: birth, expansion, removal and, now, exclusion through walls. *Biblio 3W, Revista Bibliogràfica de Geografia y Ciencias Sociales*, Universidad de Barcelona, Vol. XIV, n° 828, June 25, 2009. <http://www.ub.es/geocrit/b3w-828.htm>.

FRANÇOIS, J-C. Discontinuity. Available at: http://www.hypergeo.eu/spip.php?article15.

GLAESER, E. and KAHN, M. Sprawl and urban growth. NBER Working Paper Series, 2003.

GOMES, Paulo C. Scenarios of urban life: images, spaces and representations. Revista Cidades. v.5, n.7, 2008, p. 9-14.

HAESBAERT, Rogério. Territory, insecurity and risk in times of social contention. In: FERREIRA, A. et al (org.) A Experiência Migrante: entre deslocamentos e reconstruções. Rio de Janeiro:

Garamond, 2010, p.537-557.

LE GOFF, Jacques. For love of cities. Sao Paulo: UNESP, 1998.

MACHADO, Ana Brasil. Em cima do muro: um cenàrio para o ecolimite do Santa Marta. Monograph (BA in Geography) - Department of Geography/IGEO/UFRJ, Rio de Janeiro, 2009.

MACHADO, Ana Brasil. The wall is the news: the discourse of order and the ecolimitation of Santa Marta. Espaço Aberto Magazine, v. 1, n.2, 2011, p.157-166.

MAGALHÂES, Sérgio. About the city: housing and democracy in Rio de Janeiro. Sao Paulo: ProEditores, 2012.

MELO, Nicole. Public policy for the favelas in Rio de Janeiro: the problem (in) framing. International Institute of Social Studies, 2010.

MUMFORD, Lewis. The city in history: its origins, transformations and prospects. Sao Paulo: Martins Fontes, 2008[1982].

OLIVEIRA, Fabricio. Territorial constraints for the preparation of population estimates for sub-municipal units: considerations from the case of Rio de Janeiro. Coleçao estudos cariocas. Rio de Janeiro, 2008.

PEDROSO, Isabella. O Estado e os Muros: um estudo sobre as políticas destinadas às favelas cariocas através da mídia impressa. Rio de Janeiro, UERJ, Monograph defended in the Postgraduate Program in Social Sciences, Specialization Course in Urban Sociology, 2009.

PENDALL, R., MARTIN, J., and FULTON, W. Holding the line: urban containment in the United States. The Brookings Institution Center on Urban and Metropolitan Policy, 2002.

RAFFESTIN, Claude. Towards the social function of the border. Spaces and Societies, n. 70-71, 1992, p. 157-163.

REDE RIO CRIANÇA et al. The walls in the favelas, 2009. Available at http://global.org.br/wp-content/uploads/2009/12/Relat%C3%B3rio-Os-Muros-nas- Favelas-e-o-Processo-de-Criminaliza%C3%A7%C3%A3o.pdf.

REITEL, B. Frontière. 2004. Available at: http://www.hypergeo.eu/spip.php?article16.

REVEL, Judith. Foucault: essential concepts. Sao Paulo: Claraluz, 2005.

RIBEIRO, Leticia Parente. The twin cities of Foz do Iguaçu and Ciudad Del Este: spatial interactions on the Brazil-Paraguay border. Rio de Janeiro: UFRJ, 2001.

RIO DE JANEIRO. Decree 322 of 03/03/76. Approves the Zoning Regulations for the Municipality of Rio de Janeiro, 1976.

RIO DE JANEIRO. Complementary Law 111/2011. Sustainable Urban Development Master Plan for the Municipality of Rio de Janeiro, 2011.

RIO DE JANEIRO. Bill 245/09. Establishes regulations for the implementation of ecolimits in the Municipality of Rio de Janeiro, 2009.

ROSALES, Verónica and SANCHÉZ, Pedro. Sustentabilidad urbana: planteamientos teóricos y conceptuales. Quivera, vol. 13, n. 1, 2011, p.180-196.

SCHLEE, Mônica and ALBERNAZ, Maria Paula. Protection of hillsides by municipal legislation: an assessment of the current situation in the city of Rio de Janeiro. Florianópolis: Anais XIII ENANPUR, 2009

Annex: example of the form used in the content analysis of the documents considered

<table>
<tr><th colspan="2">Device Element Identification</th></tr>
<tr><td>Title</td><td>Illegal. So what: the expansion of favelas has no eco-limits. Audit by the Court of Auditors proves that communities are rapidly expanding into preserved areas</td></tr>
<tr><td>Document Type</td><td>Newspaper article</td></tr>
<tr><td>Producer agent</td><td>The Globe</td></tr>
<tr><td>Date of publication</td><td>16/10/2005</td></tr>
<tr><td>Other information</td><td>Editorial Rio, Primeiro Caderno, page 9</td></tr>
<tr><th colspan="2">Spatial categories mobilized</th></tr>
<tr><td rowspan="4">1</td><td>Identification: 17 favelas</td></tr>
<tr><td>Morphology: They occupy areas of environmental preservation</td></tr>
<tr><td>Uses and actions (current and prescribed):</td></tr>
<tr><td>Meanings: Slum expansion has no eco-limits (subtitle) Many put a rich ecosystem at risk.</td></tr>
<tr><td rowspan="4">2</td><td>Identification: 42 communities</td></tr>
<tr><td>Morphology: They are located at a maximum distance of one hundred meters from areas administered by the Union, the state and the municipality.</td></tr>
<tr><td>Uses and actions (current and prescribed):</td></tr>
<tr><td>Meaning:</td></tr>
<tr><td rowspan="4">3</td><td>Identification: Environmental preservation areas</td></tr>
<tr><td>Morphology: Some of these areas are located near slums/communities</td></tr>
<tr><td>Uses and actions (current and prescribed):</td></tr>
<tr><td>Meaning: Areas that are at risk and must be preserved</td></tr>
<tr><td rowspan="4">4</td><td>ID: Reservation-favelas</td></tr>
<tr><td>Morphology:</td></tr>
<tr><td>Uses and actions (current and prescribed): Transformation of parks (environmental preservation areas)</td></tr>
<tr><td>Meaning: We run the risk of these parks being transformed into what I call slum-reserves, due to a lack of control over expansions .[69]</td></tr>
</table>

69 Statement by the then president of the Legislative Assembly's Environment Committee, Carlos Minc (PT).

5	ID: ***Rocinha, in Laboriaux***
	Morphology: There are houses very close to the boundaries of Tijuca Park
	Uses and actions (current and prescribed):
	Meaning:
6	ID: ***Tijuca Park***
	Morphology: Houses in Rocinha, Laboriaux are located close to the park boundary
	Uses and actions (current and prescribed):
	Meaning:
7	**Identification:** ***Among the trees***
	Morphology: In Babylon
	Uses and actions (current and prescribed): Houses outside the eco-limits
	Meaning:
8	**Identification:** ***Green areas***
	Morphology: They are separated from the buildings by the eco-boundaries
	Uses and actions (current and prescribed):
	Meaning:
9	**ID:** ***Outside the eco-limits***
	Morphology: In Babylon
	Uses and actions (current and prescribed): *There are already 86 houses*
	Meaning:
Limits	
1	**ID: Eco-limits**
	Function: *Separate green areas from buildings*
	Physical Dimension:
	Meaning:
	Location: Babilônia
	Condition of implementation: Demarcated by the town hall
Interfaces	

<table>
<tr><td rowspan="4">1</td><td colspan="2">Related space systems:</td></tr>
<tr><td>BABYLON
GREEN AREAS</td><td>AMIDST THE TREES/ OUTSIDE THE ECO-LIMITS/</td></tr>
<tr><td colspan="2">Discontinuity: Ecolimits</td></tr>
<tr><td colspan="2">Relationship between spatial systems: Despite the establishment of eco-boundaries, buildings in Babylon occupy green areas outside the eco-boundaries.</td></tr>
<tr><td rowspan="4">2</td><td colspan="2">Related space systems:</td></tr>
<tr><td>ROCINHA, IN THE LABORIAUX</td><td>TIJUCA PARK</td></tr>
<tr><td colspan="2">Discontinuity:</td></tr>
<tr><td colspan="2">Relationship between spatial systems: Laboriaux's buildings are located close to Tijuca Park, which raises concerns about its growth over the preservation area.</td></tr>
<tr><td rowspan="4">3</td><td colspan="2">Related space systems:</td></tr>
<tr><td>17 FAVELAS / 42 COMMUNITIES</td><td>ENVIRONMENTAL PRESERVATION AREAS</td></tr>
<tr><td colspan="2">Discontinuity: Maximum distance of one hundred meters</td></tr>
<tr><td colspan="2">Relationship between spatial systems: The favelas and communities already occupy areas of environmental preservation or are located very close to them, putting rich ecosystems at risk.</td></tr>
</table>

yes
I want morebooks!

Buy your books fast and straightforward online - at one of world's fastest growing online book stores! Environmentally sound due to Print-on-Demand technologies.

Buy your books online at
www.morebooks.shop

Kaufen Sie Ihre Bücher schnell und unkompliziert online – auf einer der am schnellsten wachsenden Buchhandelsplattformen weltweit! Dank Print-On-Demand umwelt- und ressourcenschonend produzi ert.

Bücher schneller online kaufen
www.morebooks.shop

info@omniscriptum.com
www.omniscriptum.com

Printed by Books on Demand GmbH, Norderstedt / Germany